Land Tenure Security and Sustainable Development

Margaret B. Holland • Yuta J. Masuda
Brian E. Robinson
Editors

Land Tenure Security and Sustainable Development

palgrave
macmillan

Editors
Margaret B. Holland
Department of Geography &
Environmental Systems
University of Maryland, Baltimore County
Baltimore, MD, USA

Yuta J. Masuda
Global Science
The Nature Conservancy
Arlington, VA, USA

Brian E. Robinson
Department of Geography
McGill University
Montreal, QC, Canada

ISBN 978-3-030-81880-7 ISBN 978-3-030-81881-4 (eBook)
https://doi.org/10.1007/978-3-030-81881-4

This Palgrave Macmillan imprint is published by the registered company Springer Nature Switzerland AG.
The registered company address is: Gewerbestrasse 11, 6330 Cham, Switzerland

We dedicate this volume to land rights defenders around the world, both those who have come before, and those who are working now to raise the voices of local peoples.

Foreword

In the fierce competition for resources, property rights and tenure security are vital centerpieces of the global development agenda. For this to be actionable, we need a coherent articulation of relationships between tenure security and sustainable development. Land tenure underpins the socio-economic relations surrounding natural resource use and, especially, decisions around whether to increase agricultural production, decrease environmental degradation, support urban development, empower women, and support property rights for Indigenous Peoples and local communities. Without a doubt, tenure security significantly influences how different persons will respond to and participate in development processes. Whoever holds tenure controls the outcome of production, including the benefits, the flow of which is kept in check by institutions developed by and within communities. It directly determines who has the right to benefit from land-based resources and who has to protect them. Security is thus the assurance that the rights of a landholder will be respected and upheld in the event of a challenge, dispute, or risk as the competition for land and its resources ensues. It is a fallacy to ascribe security to any particular form of land tenure.

Tenure security is inextricably linked to current global sustainability agendas that merge the aspirations of humanity with dignity and the respect for human rights, especially those of persons involved in regeneration and restorative actions, seeking to reboot the health and functioning of our environmental systems, biodiversity, and ecological richness. The case studies in this book mirror my understanding of tenure security as bi-relational in character, bridging the twin role of beneficiary and steward as intertwined with people's livelihoods and the lands they depend on. I have seen this illustrated by pastoralist communities in the Karamoja region of Uganda, where tribal elders are holders of collective tenure, demarcating their grazing lands into dry pastures, shifting them from wet grazing areas during times of harsher weather conditions, to ensure continuity in the beneficial and sustainable use of their land.

Progress toward universal conditions of equity of human agency is strongly dependent on how such interests and rights over land are governed to secure tenure. Still, contradictions and conflicts between and within institutions remain, given different actors carry distinct visions of how such trade-offs should be resolved concerning resource use. Within the chapters of this book, it is clear that women, Indigenous Peoples, and local communities are at the heart of ensuring the continuous flow of ecosystem services for current generations and into the future. Sustainable development is squarely focused on promoting human agency and well-being, both within current generations and in the generations to come.

All forms of land tenure, whether private, community, state, or common, are susceptible to insecurity. Their susceptibility depends on the composition of their bundle of rights in a particular context: how rights are defined, or how they are enforced. Each context has a unique mix of bundles, definitions, and delineation—hence a diversity that sometimes creates confusion or results in divergent development paths. This book demonstrates the criticality of secure tenure to sustainable resource use and to intra- and inter-generational equity that safeguards the vast majority of the world's ecosystem services, despite the variation or binaries in the paths taken in the pursuit of development. In my capacity advising the government of Uganda on the National Land Policy 2013, it was clear that a universal or 'one-size-fits-all' path to secure tenure for all social-economic groups, such as pastoralists, women, urban dwellers, and forest dwellers, was not tenable. However, our efforts were based on shared principles to balance the role of land in the national economy, while at the same time continuing to secure livelihoods and create incentives to safeguard resources into the future.

Institutions, whether formal or informal, are devised to uphold tenure rights or property rights associated with the land, to shape human interactions, and to regulate behavior or relations through deterrents and rewards. Secure tenure has a triple dimension of legality, legitimacy, and certainty, which is essential to realizing the benefits within a bundle of rights, mediating investment, safeguarding incentives, securing inclusion in governance, and guaranteeing the safety of property over land. Legal security is based on rights established and defended in law, as protections or restrictions, conferring rights and requiring action, such as preventing arbitrary loss of rights. Legitimacy is inextricably linked to the common or popular acceptance of a practice within a system, whether social, cultural, governance, or leadership. Perceived security is situational certainty or uncertainty linked to the presence or absence of a condition, based on the soft concept of conscious observation.

These three strands of tenure security scrutinized in this book reveal their interrelations in the pursuit of development agendas. For instance, giving legal title to a property in a manner that recognizes possessory claims has been linked to economic development and poverty reduction in much of the developed world. However, tenure security wields greater relevance and importance to the problems of the poor in Africa and Asia, and perhaps to a lesser extent in South America. This variance in assumptions that govern the definition, delineation, and enforcement of property rights has delivered a different reality in Africa where rights are collective or customary, as the favor is bestowed upon formal, individual, and registered tenure, to support the functionality of the market, even as this sometimes lacks legitimacy within society or amongst communities. This results in binaries of economic formality versus informality (the dual economy). Indeed, sustainable development that is inclusive and equitable has to embrace the flexibility, diversity, duality, or plurality that arises with the deployment of tenure security for individual and collective, state, and private entities, given the bi-directional and multi-directional nature of rights over land, as well as the varied meanings of land rights to people and societies.

What is of essence is an appreciation of what exists in tenure systems that are socially entrenched and laden with economic potential to fit existing political circumstances. This helps expand the recognition of property rights to spur, support, and sustain progress consistent with the aspirations of the 2030 Development Agenda. It is essential to keep the enduring forms of property relations and land institutions to guarantee the dividends of tenure security, accepting the diversity of tenure regimes without losing sight of the basic functions and value of resources on land is at the heart of sustainable development. In this book, approaches to strengthening tenure security are discussed, vividly demonstrating the blend of social, cultural, spiritual, natural, and economic aspects of tenure security as a complex relational reality, constituted in co-existence of enduring norms with formal rules and the highest level of acceptance of diversity in development paths and agendas.

Landesa
Dar es Salaam
Tanzania

Margaret A. Rugadya

Preface

Land tenure security has come to the forefront of the sustainable development agenda in recent years. In part this is due to its foundational and fundamental nature. Ways to manage and allocate rights over resources, and our relationships to it, are some of the first 'commons' issues fledgling societies face. The ripple effects of land tenure security are now widely recognized as having implications for not just the way natural resources are managed, but also for household income and investment, well-being, and health.

This book came out of discussions from a series of working group meetings convened by Maggie, Yuta, and Brian (the editors of this book). These workshops focused on the role of land tenure security in conservation activities and programs, but as a group we quickly recognized conservation was just one topic that land tenure security affects. The working group met three times over two years, and several participants in the working group are authors on chapters herein. In the final meeting, the group discussed critical resources needed for practitioners, students, policymakers, and researchers interested in land tenure security and its role in advancing sustainable development, and whether these resources existed. The group converged on the idea that a book project could provide a space for grounded, approachable discussions on the various topics land tenure security affects.

Why did we think a book was needed? There are several reasons we thought a book like this would fill a gap in what is already volumes written on land tenure (and tenure security). First, little exists that links land tenure security across a range of topics and conveys how land tenure has fundamental links to so many important social and environmental outcomes. We also recognize that land tenure is complex but is gaining wider attention, so we hoped to examine these complexities in ways that are easily accessible to a general audience. For readers that are unfamiliar or new to tenure security, the topic can be overwhelming, complicated, and costly to study. We believe this is a primary barrier for many that are eager to understand the nuances around tenure security, but lack the time, expertise, or assistance to do so. We hope this book can serve as an introduction to issues around land rights and how these fit into the sustainable development agenda, but also be rich with real-world cases that showcase fundamental topics and issues.

Land and resource rights are central to many conversations that have been emerging in the past several decades. This includes taking steps to reconcile deep imprints of colonial histories and recognize rights of Indigenous, marginalized, and vulnerable communities. These conversations have emerged in low-, middle-, and high-income regions. We also recognize that many of the authors of chapters in this book are working from universities and institutions that are often founded on unceded territories, that much of our academic training is embedded in an exploitive and systemically racist society, and that academic reputations are sometimes built on knowledge and examples which are not their own. When soliciting authors to contribute chapters, we had an explicit goal for each chapter to not just draw from leading academic and theoretical perspectives around land tenure security, but to also include and recognize voices from practitioners and those embedded in the programmatic side of making land rights secure for people around the world. This is perhaps a small step toward addressing the large issues we face in developing a more just and inclusive society, but our greatest hope is that perhaps these chapters collectively elevate issues and voices and spark conversations that help address current and historical injustices.

This book is organized into four sections. In Chaps. 1, 2, and 3, we lay out some foundational and introductory information. Our goal for this section is for it to serve as a primer for foundational concepts and history tied to land tenure security. Chapters 4, 5, 6, 7, 8, 9, and 10 are more empirically focused, presenting cases and details on how land tenure security underpins numerous sustainability challenges that face us today. Chapters 11, 12, and 13 describe several approaches to addressing land tenure security issues. The final section of the book, Chaps. 14 and 15, close with robust methods for assessing and researching land tenure security, as well as thoughts about ways forward for integrating research and policy.

As we continue to strive for targets laid out by the Sustainable Development Goals, and as we develop new international frameworks and agendas to pursue, the security of land and resource rights will remain key challenges faced by many communities. We hope these chapters help increase the awareness of key issues and challenges, and we look forward to continuing these conversations with communities, academics, NGOs, multilateral institutions, and others as we move forward.

Margaret B. Holland
Baltimore, MD, USA

Yuta J. Masuda
Arlington, VA, USA

Brian E. Robinson
Montreal, QC, Canada

Acknowledgments

We are grateful to the generous financial support provided by the David and Lucile Packard Foundation (Grant/Award Number: 2016-64676), The Nature Conservancy, the Social Sciences and Humanities Research Council of Canada (Grant/Award Number: 244615), and the McGill Sustainability Systems Initiative. Their support allowed us to convene a diverse group of experts to cultivate many of the ideas that emerge in this book and provided the flexibility needed to work on this project. We are also deeply grateful to the many practitioners, community members, and others who took the time to provide thoughtful, detailed, and interesting stories that ground many of the concepts in this book.

Contents

Notes on Contributors

Rodrigo Arriagada is an environmental economist focusing on ecosystem services and impact evaluation tied to environmental policy and governance. He holds a PhD in Environment and Resource Economics from North Carolina State University and the geographic focus of his work has been oriented within Latin America. Dr. Arriagada is an Associate Professor and Director of Environment and Sustainable Development at the Pontificia Universidad Católica de Chile.

Allen Blackman is an environmental economist and Principal Economic Advisor in the Inter-American Development Bank's Climate and Sustainable Development Sector. His research focuses on land use/cover change, agricultural supply chains, and impact evaluation of conservation-oriented policies. He holds a PhD in Economics from University of Texas-Austin and worked for many years with Resources for the Future.

Brianna Castro is a doctoral candidate in Sociology at Harvard University studying the social impacts of climate change with specializations in climate displacement and slow-onset environmental stressors such as drought, coastal erosion, and nuisance flooding. She holds a BA in Public Policy and Spanish Language and Literature, served in the US Peace Corps in Colombia, and has collaborated with organizations such as UN Women, Sierra Club, and The Nature Conservancy.

Malcolm Childress is a multidisciplinary land resources specialist with 30 years of global experience, including on urban and rural property rights, as well as developing strategies for managing critical global ecosystems. His focus areas include land policy and governance, land markets, land registration, property taxation, cadastral systems, and spatial planning. Dr. Childress is the Executive Director with the Global Land Alliance.

Pranab Choudhury is a practicing researcher and consultant in the areas of natural resources management and governance, with two decades of experience. He works on inter-disciplinary issues around forest, land, water, and livelihoods in India, Nepal, and Bangladesh. Mr. Choudhury is a Vice President for NRMC at the Center for Land Governance in India.

Shipra Deo leads Landesa's work to strengthen women's land rights in India. She is a development practitioner with more than 20 years of experience in managing multidisciplinary programs addressing women's empowerment, women's land rights, violence against women, agriculture, collective action, livelihoods, and institution building.

Moustapha Diop is a land tenure specialist with nearly 15 years of experience working on land tenure issues and property rights in Africa. He holds a PhD in Geography from the University of Maryland Baltimore County (UMBC), and his research focuses on the intersections between tenure security, food security, and sustainable agriculture in West Africa.

Diana Fletschner is the Chief Program Officer at Landesa. She is a development economist with over 25 years of experience teaching, conducting research, designing programs and tools, monitoring and evaluating interventions, mentoring, and conducting strategic advocacy with a special focus on rural women. She has co-led sector-wide efforts that resulted in the successful inclusion of land rights among Sustainable Development Goals' targets and indicators, creating both an ambitious commitment to ensure land rights for all and a global mandate to generate land rights data.

Lea Fortmann is Associate Professor of Economics at the University of Puget Sound. Her primary research focuses on the role of community forest management in reducing deforestation and promoting sustainable livelihoods with a particular focus on Guatemala's Maya Biosphere Reserve. She holds a PhD from The Ohio State University and an MPA from the University of Washington.

Alain Frechette is the Director of Strategic Analysis and Global Engagement at Rights and Resources Initiative. He has over 25 years of experience in natural resource management, biodiversity conservation, and climate change. He began his career with state and provincial forest and protected area agencies in the US and Canada before pursuing strategic consultancies with multilateral organizations, development agencies, and NGOs such as IUCN, DFID, the World Bank, and various UN institutions across Africa, Asia, Europe, and Latin America.

Fernando Galeana is Assistant Professor of Sociology and Integrative Conservation at William and Mary. He specializes in political ecology, environmental justice, and development studies. His professional experience includes working with multilateral organizations, NGOs, and Indigenous organizations in the defense of collective land rights.

Raoul Herrmann is an Economist and Associate Researcher at the German Development Institute. His research interests concern sustainable economic and social development, including large-scale land investments.

Margaret B. Holland is an Associate Professor at the University of Maryland, Baltimore County (UMBC). She is an applied human-environment geographer whose work explores the intersections between land tenure, rural communities, land use dynamics, and conservation strategies with a focus on forests and protected areas. She has active research in Maryland, Central America, Ecuador, and Mozambique. Maggie holds a PhD in Environment & Resources from the Nelson Institute at the University of Wisconsin-Madison.

Sarah Holzapfel is a former Senior Researcher at the German Development Institute. Her work has focused on how to increase the results orientation of agricultural, rural development, and food security interventions in line with the aid effectiveness agenda.

Kelly W. Jones is an Associate Professor at Colorado State University. She applies microeconomic theory and methods to understand relationships between conservation governance, land cover change, ecosystem services, and human well-being. She holds a PhD in Agricultural and Natural Resource Economics from the University of Wisconsin-Madison.

Christina Kuntz has over ten years of experience working in international development including as a Peace Corps volunteer, program director, and researcher. Based in Colombia, she holds a BA in Psychology and an MA in International Development and works with the Global Land Alliance and USAID on land tenure research in Colombia.

Steven Lawry holds a PhD from the Nelson Institute of Environmental Studies at the University of Wisconsin-Madison. He is a Senior Associate at the Center for International Forestry Research and has worked for the Land Tenure Center at the University of Wisconsin-Madison and the Ford Foundation. His research focuses on farmer and community responses to land tenure interventions and on the conditions that enhance community stewardship of forests and other natural resources.

Edward Lekaita is an advocate of High Court of Tanzania and Subordinate Courts thereto save for Primary Courts. He has been a legal advisor, and Head of Advocacy and Coordinator for Kiteto Program. He has worked for the Ujamaa Community Resource Team since 2010 mainly on policies, laws on natural resources and governance, land use planning, certificates of customary rights of occupancy, and wildlife management areas for over ten years.

Henry Machina is a gender, land, and youth specialist whose work has focused on land policy, youth in agriculture, and women in agribusiness in Zambia. He is affiliated with the Indaba Agricultural Policy Research Institute (IAPRI) and is the director of the Zambia Land Alliance.

Monica Mhoja has more than 25 years of experience in managing and leading women's and children's rights programs. She has been designing, managing, and implementing community, national, regional, and international programs on advocacy, legal and human rights awareness on land rights, inheritance rights, and the impact of child marriage and gender-based violence using collaborative and human rights-based approaches. Dr. Mhoja works at Landesa as the Outreach Director for Africa.

Jackson Marubu is the Kenya Program Coordinator at The Nature Conservancy, managing various projects under the country program with a wide array of stakeholders in the Kenyan government, multinational organizations, and global conservation teams. Prior to his current role, he worked as a technical officer at Freeland-Africa managing a small team working with wildlife law enforcement agencies across the region, with the vision of a world free of wildlife trafficking and human slavery.

Yuta J. Masuda is a Senior Sustainable Development and Behavioral Scientist in Global Science at The Nature Conservancy. His work at the Conservancy investigates the impacts of conservation programs on human well-being, and he has a particular interest in land tenure, gender, development, institutions, and human health. His work looks at integrating human well-being considerations into conservation programs to better understand their risks and benefits to people.

Roan McNab is the Guatemala Program Director for the Wildlife Conservation Society based out of their Flores, Petén field office. He holds an undergraduate degree in Anthropology from Duke University and master's degree in Tropical Conservation and Development from the University of Florida's Center for Latin American Studies.

Daniela A. Miteva is an applied microeconomist working primarily on sustainability issues in developing countries. She is an Associate Professor in the Department of Agricultural, Environmental, and Development Economics at The Ohio State University in Columbus, OH (USA).

Iliana Monterroso is a scientist and co-coordinator of Gender and Social Inclusion Research at the Center for International Forestry Research. In 2022 she joined the Climate and Land Use Alliance (CLUA) as Lead Consultant for Mexico and Central America. Her research focuses on gender, tenure, collective rights, environmental governance, and socio-environmental conflicts predominantly in Latin America.

Patricia Mupeta-Muyamwa is a natural resource governance specialist and the Africa Indigenous Landscapes Strategy Director at The Nature Conservancy. She has 20 years of experience in protected area management, community conservation, and rural development. She has worked for TNC since 2011. She leads the Indigenous Landscape program, which supports communities in eastern and southern Africa with stronger land and resource rights, builds their leadership capacity to manage landscapes, and creates economic opportunities for improved livelihoods.

Chewe Nkonde is a lecturer and a researcher in the Department of Agricultural Economics and Extension at the University of Zambia. His research covers the following thematic areas: agricultural productivity, food and nutrition security, food and agricultural markets, land distribution and farm structure change, and issues at the nexus of agriculture, natural resource sustainability, and economic development.

Kerstin Nolte is an economic geographer holding a PhD in Economics from the University of Göttingen, Germany. She is an Assistant Professor at the Institute of Economic and Cultural Geography at Leibniz University Hannover, Germany. Her research focuses on sustainable rural transformation with a focus on large-scale land acquisitions. Her research on the processes in which land acquisitions take place, on the determinants and impacts, builds on field research in Mali, Zambia, and Kenya.

Daudi Peterson has supported and facilitated Indigenous land rights and community-based conservation in Northern Tanzania for the last 30 years. He is a co-founder and current board member of the Dorobo Fund, Dorobo Safaris, and the Ujamaa Community Resource Team (UCRT) in Tanzania.

Brian E. Robinson is an Associate Professor at McGill University. He has published widely on the role of property rights in conservation and resource management. His research looks at policies that mediate the relationship between humans and the environment, how people meet their needs through the use of ecosystems and natural resources, and the role these issues play in human well-being. His methods draw from environmental and development economics, development geography, ecology, and land systems science.

Margaret Rugadya holds more than two decades of experience as a development practitioner and academic research, and she is a leader in the area of land policy and gender equity, with contributions to the economics of gender, land, and resource tenure. Rugadya holds a PhD in Economics—Governance and Public Policy from Maastricht University. The focus of her work is on women, pastoralists, forest dwellers, indigenous persons, and urban dwellers across Africa.

Paul Samboko is an agricultural economist whose research interests span the development environment nexus in sub-Saharan Africa. His previous work involved capacity building, evidence generation, and outreach activities relevant to agriculture in Zambia and other countries in Eastern and Southern Africa.

Jolyne Sanjak is an agricultural economist with specialization in development economics. She has more than 25 years of technical, research, and managerial experience relating to inclusive global economic development, with technical expertise in areas including rural and urban land governance, as well as with rural livelihoods and agricultural development. Dr. Sanjak is a Sector Director for Land Tenure and Property Rights, with Tetra Tech, ARD.

Laura Aileen Sauls is a Leverhulme Early Career Fellow in the Department of Geography at the University of Sheffield and an affiliated research scientist with the Clark University Center for the Study of Natural Resource Extraction and Society. Her research primarily concerns movements for Indigenous and community forest rights in the context of global political and environmental change. In late 2022, she starts as an Assistant Professor of Global Affairs at George Mason University in Virginia.

Chira Schouten is a natural resource manager with over 25 years of experience in project management, organizational management, and strengthening networks that involve multiple stakeholders in both the conservation and agriculture sectors. She has worked for The Nature Conservancy since 2013 first as the program lead for northern Tanzania and now manages the Africa conservation portfolio consisting of four regional strategies and nine program teams.

Mourad Shalaby is a Project Evaluation Analyst at the Global Environment Facility. He is a multilingual climate change expert with over ten years of professional experience with several prominent international organizations and research institutes in the environmental field with expertise in environmental and social impact assessments. He holds masters' degrees in Environmental Geography and Urban Planning from McGill University and Université de Montréal, respectively.

Erin Sills is a Professor and Head of the Department of Forestry and Environmental Resources at North Carolina State University, as well as a long-time associate of the Center for International Forestry Research. Her research lies at the interface of environment and development, including rigorous evaluation of the local impacts of conservation interventions across the tropics.

Mitelo Subakanya is a Research Associate at the Indaba Agricultural Policy Research Institute (IAPRI). Her research interests include environment and natural resources management, climate change, poverty analysis, impact evaluation, agriculture, and rural development and rural financing.

William D. Sunderlin has conducted research on the underlying causes of tropical deforestation, forests and livelihoods, forests and climate change (specifically on REDD+), and forest tenure. He worked at the Center for International Forestry Research (CIFOR) in Bogor, Indonesia, in the periods 1994–2006 and 2009–2015 and at the Rights and Resources Initiative (RRI) in Washington, DC, 2006–2009. He is now a Senior Associate (CIFOR), a Fellow (RRI), and a Visiting Professor at the State University of New York—College of Environmental Science and Forestry.

Hannah L. Timmins has over seven years of experience developing technical expertise in conservation area best management practices, expansion, restoration, and connectivity. She has managed and led conservation and research projects and developed publications for organizations including Zoological Society of London, The Nature Conservancy, International Union for Conservation of Nature, Convention on

Biological Diversity, Forest Conservation Fund, and BirdLife International.

Tzu-Wei Joy Tseng is an epidemiologist and works on a variety of topics related to human health and social policy. She has experience assessing how land tenure security affects conservation and human well-being outcomes and has also investigated the link between household air pollution and health in China. She is a graduate researcher with the Institute for Health and Social Policy at McGill University in Montreal, Canada.

Jon Unruh is a Professor at McGill University, and his work focuses on war-affected land and property rights. He has conducted research, policy and practice work in Syria, Yemen, Lebanon, Sudan, Liberia, Somalia, Mozambique, East Timor, Sri Lanka, Sierra Leone, Zambia, Ethiopia, Uganda, Cambodia, Angola, Cameroon, Turkey, Slovenia, Jordan, Colombia, Peru, Tanzania, Kenya, and Saudi Arabia and generally with the UN, World Bank, USAID, DFID, and other bilateral donors and NGOs.

Abbreviations

AfD	Agence Française de Développement
BACI	Before-after-control-impact
CAR	Corrective action request
CATIE	Center for Tropical Agriculture and Education
CBC	Community-based conservation
CCROs	Certificates of Customary Right of Occupancy
DAG	Directed acyclic graph
DGM	Dedicated Grant Mechanism
DID	Difference-in-differences
ES	Ecosystem Services
FAO	Food and Agriculture Organization of the United Nations
FMUs	Forest Management Units
FSC	Forest Stewardship Council
GIZ	Deutsche Gesellschaft für Internationale Zusammenarbeit
IBA	Important Bird Area
IPLCs	Indigenous Peoples and local communities
IV	Instrumental variables
KWCA	Kenya Wildlife Conservancies Association
LTS	Land tenure security
MBR	Maya Biosphere Reserve
MCC	Millennium Challenge Corporation
MMWCA	Maasai Mara Wildlife Conservancies Association
NGO	Non-governmental organization

NTRI	Northern Tanzania Rangelands Initiative
PES	Payment for ecosystem services
PSM	Propensity score matching
RDD	Regression discontinuity design
REDD+	Reducing Emissions from Deforestation and forest Degradation—plus
SCM	Synthetic control method
SDGs	Sustainable Development Goals
SIDA	Swedish Institute for Development Assistance
TNC	The Nature Conservancy
TOC	Theory of change
USAID	United States Agency for International Development
VIA	Value added impact
WWF	World Wildlife Foundation

List of Figures

List of Tables

List of Boxes

1

Property Rights, Tenure Form, and Tenure Security

Yuta J. Masuda, Brian E. Robinson, and Margaret B. Holland

While trekking through the forests of Indonesia Borneo far from any urban center, the sound of insects, mammals, reptiles, and the roaring river can be deafening. It is a testament to the forest's immense biodiversity—some of which is found nowhere else in the world. Yet no more than a few kilometers away, the forests open up to an expanse of oil palm trees that stretch to the horizon, neatly planted in rows like corn fields in central Indiana. This stark contrast in landscapes masks the tension around land ownership and land use in Indonesia Borneo that has lasted for several decades. Market forces fueled by policies that catalyzed demand for oil palm led to targeted increases in the production of oil palm. This

Y. J. Masuda (✉)
Global Science, The Nature Conservancy, Arlington, VA, USA
e-mail: ymasuda@tnc.org

B. E. Robinson
Department of Geography, McGill University, Montreal, QC, Canada

M. B. Holland
Department of Geography & Environmental Systems, University of Maryland, Baltimore County, Baltimore, MD, USA

M. Holland et al. (eds.), *Land Tenure Security and Sustainable Development*,
https://doi.org/10.1007/978-3-030-81881-4_1

led the way for international conglomerates to purchase hundreds of thousands of acres for oil palm production. But the lands being targeted for oil palm production were oftentimes occupied by local Dayak communities and others who have lived in the forests of Indonesia Borneo for generations, many of which settled the area there long before land titles. Indeed, in the 1980s, the Indonesian government recognized longstanding settlements and arranged systems whereby villagers could arrange a profit-sharing agreement with organizations interested in cultivating and developing the land in exchange for transferring development rights. But as an extensive *New York Times* piece (Lustgarten, 2018) documented, "companies often secured the permits they needed through some combination of intense lobbying, bribery and strong-arming, and the result was broken promises and missing payments." Villagers lacked the resources or institutional knowledge of how to defend their rights. Due in part to the power differential between villagers and international conglomerates and the enormous demand for oil palm, 16,000 square miles of rainforest have been lost since 1973, which accounts for approximately 20% of deforestation in impacted areas (Gaveau et al., 2016). The overall socio-economic impact of oil palm expansion remains difficult to generalize (Sheil et al., 2009), but increasing evidence suggests land ownership and land use remains central to concerns, conflict, and debate about increasing oil palm production and who ultimately benefits (Rist et al., 2010).

On the other side of the world in the Middle Belt region of Nigeria, National Public Radio (McDonnell, 2018a) reported a vicious cycle of retaliatory attacks between permanent farming communities and the nomadic pastoralist Fula that has persisted for years. Amnesty International (2018) suggests a primary cause stems from disputes over access and ownership over water, land, and pasture, resulting in over 3600 people killed from January 2016 to October 2018. Recent shifts by farmers to dry season farming, population growth, and shifting livelihood strategies by both herders and farmers increased demand for natural resources. Water, for instance, was needed by farmers for irrigation and by herders for their livestock in the same season. Prior to these changes, traditional agreements between farmers and herders largely prevented conflict, and when conflicts did occur, they were quickly resolved. Individual herders lacked land rights and instead utilized communal lands, and herders also received permission to graze livestock in areas that were not being used for agriculture by farmers. In response to increasing conflicts, some state governments instituted restricted grazing to enclosed ranchland (McDonnell, 2018b). This did little to stem conflict, however, as many herders were unable to comply because they relied on communal lands that were not enclosed and also did not own any land. In return, the policy has been attributed with an increase in fatalities as, "The grazing laws pushed people to a level where they really felt like they had to fight, especially in rural areas where there is no presence of security" (McDonnell, 2018a).

A Common Thread Underpins Pressing Sustainability Challenges

The cases above illustrate how land tenure security (herein just tenure security) is an important factor for sustainable development. In these cases, tenure insecurity is both a primary and underlying issue exacerbating other pressures that create conflict and uncertainty for sustainable natural resource use or equitable development. In both cases, clarifying land tenure is critical for resolving a diverse set of issues. The cases are by no means representative of the range of issues where tenure insecurity is either a primary or a tertiary exacerbating factor. But they highlight how tenure insecurity can intersect and is driven by historical inequities, political power and influence, population growth, economic development, and other factors. Ultimately, tenure security is important for resolving issues at local, regional, and global scales, although this may not always be immediately clear.

Take, for example, the case of Indonesia Borneo, which highlights how insecure tenure has created conditions for unsustainable and arguably inequitable development of the region. The cascading effects from insecure tenure have implications for climate change, biodiversity, and sustainable economic development at all scales. On the local scale, Indonesia Borneo's forests are home to historic Dayak tribes and hold immense biodiversity. Borneo is one of the richest biodiversity regions in the world, and it is known as a global diversity "hotspot." Beyond being home to the well-known orangutan, it is home to over 14,000 plant species, among which an estimated 28% are found nowhere else in the world (Roos et al., 2004). Dayak tribes have called Borneo home for hundreds of years, and the land, the forest, and everything on and below it hold cultural and spiritual significance and also play a critical role in daily life. For instance, an estimated 34% in the rural areas of Kalimantan reported forests as a source of traditional medicine to treat kidney disease, malaria, fever, and digestive problems (Abram et al., 2014). At the regional scale, insecure land tenure may have contributed to slashing and burning of peatland and forests (Lustgarten, 2018), which in just one year is estimated to have contributed to poor air quality and led to 100,000 premature deaths in Indonesia, Malaysia, and Singapore (Koplitz et al., 2016). On the global scale, forests and peatlands in Indonesia Borneo store significant amounts of carbon, and these lands are critical for meeting ambitious global climate change goals. Here again, fires can severely setback important climate change targets. To put this into perspective, the 1997 peat and forest fires contributed to the largest increase in carbon emissions since records began (Page et al., 2002), and the 2015 fires in Indonesia Borneo released so much carbon that the daily carbon emissions were more than that of the entire European Union (Huijnen et al., 2016).

Tenure security can affect a diverse set of issues at multiple scales, so it is perhaps unsurprising that it has grabbed the attention of conservationists, ecologists, climate scientists, women's empowerment advocates, food security specialists, public health practitioners, and others. Tenure security is foundational for many global agendas, such as the Sustainable Development Goals, the Paris Agreement, and the Aichi Targets. As a result, tenure security is an explicit component of many policies and programs, such as Reducing Emissions from Deforestation and Degradation plus programs, establishment of protected areas, gender equity programs, and even sustainable agricultural programs that aim to decrease fertilizer runoff.

The interest of multiple stakeholders has led to significant growth in research on tenure security and has created a rich body of knowledge (Fig. 1.1). But it has also created fragmented research agendas and

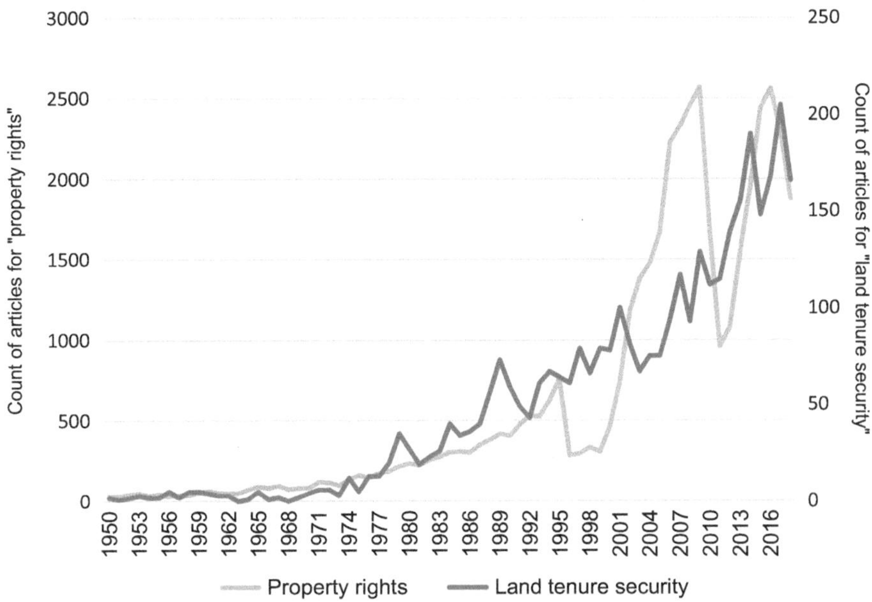

Fig. 1.1 Number of research articles for property rights and tenure security across time. Data were extracted from Scopus using the following search strings. For *property rights*, "(TITLE-ABS-KEY(("land" OR "lands") AND (("ownership right*" OR "property ownership" OR "property right*") OR ("property titl*" OR "titl*") OR ("land right*" OR "right*")) AND NOT "urban") AND PUBYEAR > 1949 AND (LIMIT-TO (DOCTYPE,"ar")) AND (LIMIT-TO (LANGUAGE,"English")))." For *land tenure security*, "Your query: (TITLE-ABS-KEY(("land" OR "lands") AND (("tenure secur*" OR "tenure insecur*" OR "secure tenure" OR "insecur* tenure" OR "land tenure")) AND NOT "urban") AND PUBYEAR > 1949 AND (LIMIT-TO (DOCTYPE,"ar")) AND (LIMIT-TO (LANGUAGE,"English")))"

approaches to resolving land tenure insecurity. Fragmentation has occurred across disciplines (e.g., economics, geography, political economy) and policy interests (e.g., nature conservation, women's empowerment). A lack of clarity in terminology (Schlager & Ostrom, 1992; van Gelder, 2010) has also created significant challenges (Masuda et al. 2020), such as inconsistent measurement, well-intentioned but poorly designed policies, and a multiplicity of theoretical frameworks. Given the complexity of tenure security, precise language and terminology is important for facilitating discussion and advancing knowledge.

Here, we provide descriptions and definitions of *property rights*, *tenure form*, *institutions*, and *tenure security* to facilitate consistent discussion on tenure security. These are all concepts that are important for understanding tenure security (Arnot et al., 2011; Robinson et al., 2018; van Gelder, 2010). Here, *rights* are "particular actions that are authorized," and a *property right* is "the authority to undertake particular actions related to a specific domain" (p. 250, Schlager & Ostrom, 1992). *Property rights* include access, withdrawal, management, exclusion, alienation, and due process and compensation (Table 1.1). For example, often implicit in home ownership is the right to possess the property, control the property, make modifications, and exclude others from the land on which a home sits, among others. However, in (for example) the case of Sweden, the codified principle of Allmansrätten grants individuals the right of public access. Property rights also imply the duty to not exercise a right, such as a restriction on one's own action as in the case of conservation easements on private land.

Tenure form "determine[s] who can use what resources, for how long, and under what conditions" (FAO, 2002). *Institutions* are "the humanly devised constraints that structure political, economic and social

Table 1.1 Definition of rights[a]

Right bundle	Definition
Access	Access rights allow a community and its members to enter an area.
Withdrawal	Withdrawal rights are the right to benefit from land, for subsistence or commercial purposes.
Management	Management can be defined by the legal limits of other rights, and it can also be used to empower a community to articulate its rights to alienation or the exclusion of particular resources.
Exclusion	Exclusion is the ability to refuse another individual, group, or entity access to and use of a particular resource.
Alienation	Alienation is the right to alienate one's property or the right to transfer one's rights to another entity. Basically this is the right to sell or subdivide land resources.
Due Process and Compensation	Due Process and Compensation is the right to due process and compensation in cases of eminent domain.

[a]Rights adapted from Rights and Resources Initiative (2016) and Schlager and Ostrom (1992)

interaction. They consist of both informal constraints (e.g., sanctions, taboos, customs, traditions, and codes of conduct), and formal rules (e.g., constitutions, laws, property rights)" (North, 1991). In essence, they are the "rules of the game" (North, 1990). Finally, *land tenure security* is a landholder's perception that rights will be upheld by society (Sjaastad & Bromley, 2000). As a result, tenure security results from an interaction between these concepts, and is multidimensional. Below, we use these terms to provide a better understanding through a case study.

Maasai Communities Around the Foothills of Mount Kenya

In southwest Meru County at the foothills of Mount Kenya, several Maasai villages have lived communally for generations. The Maasai pastoralist culture has had to adapt to colonialism, economic and social development, globalization, and other forces. This shift has meant that recent generations have adopted a more semi-pastoralist way of life, having permanent settlements with some small-scale agriculture, all while maintaining pastoralist traditions. Cattle provide the main form of wealth for these communities, and a collectivist culture whose identity is centered on cattle and pastoralism still dictates individual behavior and community governance. Communal land for livestock grazing continues to provide an important resource for communities.

Agricultural plots farmed by the Maasai contain many individual-level *property rights*. Property boundaries are established, giving a household individual rights to access the plot, withdraw products grown there, manage the plot as the household sees fit, exclude others from trespassing, and divide and sell the plot if they so desire. This is typically seen as the "full" bundle of rights that come along with the *tenure form private property* (Table 1.1).

The Maasai communities also retain rights to designated communal grazing land. In the case of Maasai villages in southwest Meru County, several villages graze livestock on communal land. Cattle and other livestock move around these lands to take advantage of seasonality and variation in the production of grasses, which are largely driven by rainfall patterns. The Maasai also have many *rights* associated with these lands. At the individual level, each person has access and withdrawal rights on their designated land. At the community level, there is a collective right to management and exclusion of others from enjoying the benefits of the land. However, the right of alienation does not exist either individually or collectively. In 2016, the Kenyan government formally recognized community land via the Community Land Act, designating the *tenure form* of these grasslands as community or *communal land*.

Here, property rights are enforced by both informal and formal *institutions*. First, the informal institution relies on traditional village governance structures within the Maasai communities and is made up of village elders who have advanced through traditional Maasai warrior culture. This system consists of older villagers who have been initiated as warriors, practiced as young elders, and have since graduated to becoming village elders. Warriors are grouped by age-sets—or cohorts—and are initiated into adulthood at the same time. They form close bonds and perform community duties such as protecting cattle and community members. Decisions involve all village elders through discussion about the infraction and appropriate penalty, and expectations are established via long-held social and cultural norms, and warriors are often tasked with enforcing punishments if needed. In this way, grassland tenure is socially upheld from within the community and with tacit or explicit support from other surrounding communities. The formal institution here is the state, as it bestows and enforces property rights through the Kenyan Community Land Act. Both these institutions can work together to provide assurances that society (either socially, at the local level, or statutorily, at the governmental level) will uphold these rights. Yet the institutions at the local level and the ones provided by the state can be unequal, and the Maasai communities largely use these informal or socially-upheld institutions to give landholders longer-term assurance over their rights, which provides *tenure security*.

Societal change, economic development, climate change, demographic change, changing cultural preferences, and other forces will likely test the resilience of property rights, the institutions that uphold and enforce them, and the Maasai communities' tenure security. For instance, unpredictable rainfall patterns have, in recent years, caused drought in neighboring pastoralist communities, resulting in decreased grass production. This threatened the health of cattle and other livestock, and as a result neighboring communities encroached on Maasai communal land as they searched for healthier pastures to feed their livestock. This event tested the strength of the Maasai community's informal institution, as it only provide tenure security as long as it is respected and adhered to by those within communal land boundaries. Recent migrants who were unfamiliar with and may not have respected or adhered to the existing informal institution can create conflict by breaking the existing social contract. In practice, elders from both the Maasai communities and the encroaching communities met to resolve the dispute, although skirmishes still occurred as information about the agreement took time to disseminate across the community. This event highlights how, in some ways, tenure is broadly secure through informal institutions and federally recognized land, and in theory there is an arbitration system that can deal with land disputes. In other ways, there are numerous ways in which tenure is insecure, which can prevent landholders from making land management decisions that involve long-term strategies or land investments.

Definitions Affect How We Analyze Issues and Develop Solutions

As demonstrated above, tenure security is complex, and the way in which we talk about the factors affecting it is critical for developing a consistent understanding of its drivers, its effects, and ways to resolve tenure insecurity. Providing clarity in descriptions and definitions is a first step, but the real world presents complex situations where understanding, for instance, how property rights intersect with the enforcement of property rights by informal and formal institutions is more than just acknowledging these are linked concepts. Further, the cases highlight how tenure security is connected to seemingly unrelated or distant issues. The primary effects of drought in a neighboring community (i.e., decreased biomass from a lack of rain impacting cattle and livestock health) may have been exacerbated by tenure insecurity in the Maasai communities. This is because droughts also affected neighboring pastoralist communities, and searching for more grazing land, these communities encroached on Maasai village communal land, thus creating greater land use pressure that expedited the degradation of communal grasslands. With stronger institutions that allow enforcement of property rights to exclude outside community members, the effects of the drought may have been minimized.

The following chapters use the terms outlined above to ensure a consistent discussion about tenure security and sustainable development. This book, collectively, is about showcasing how land tenure and secure rights over land are fundamental to many of the most basic tenants of human well-being and environmental sustainability. We hope to help illuminate the foundational role that tenure security plays in development and how it connects to many of the contemporary issues we struggle with today.

Starting from fundamental tenants and taking a long view of history, Chaps. 1, 2 and 3 aim to show the broad societal-level factors and determinants of tenure security. This comes through a review of terms with illustrative examples (this chapter), a view of the historical evolution of land rights (Chap. 2), and a perspective on how legal and customary rights are both needed for tenure security (Chap. 3). The core of this book, Chaps. 4, 5, 6, 7, 8, 9 and 10, focus on how tenure security interacts with contemporary topical issues, while Chaps. 11, 12 and 13 summarize potential ways to address tenure insecurity. In these chapters, in some cases tenure security may affect the Chapter topic itself (e.g., insecure land tenure can create conflict). In other cases, tenure security might mitigate how a particular issue affects social and conservation outcomes of interest (e.g., how tenure security interacts with conservation programs). Finally, Chaps. 14 and 15 synthesize across these issues to discuss new and emerging directions for research and practice.

References

Abram, N. K., Meijaard, E., Ancrenaz, M., Runting, R. K., Wells, J. A., Gaveau, D., et al. (2014). Spatially explicit perceptions of ecosystem services and land cover change in forested regions of Borneo. *Ecosystem Services, 7*, 116–127. https://doi.org/10.1016/j.ecoser.2013.11.004

Arnot, C. D., Luckert, M. K. K., & Boxall, P. C. C. (2011). What is tenure security? Conceptual implications for empirical analysis. *Land Economics, 87*(2), 297–331.

FAO. (2002). What is land tenure? In M. Cox (Ed.), *Land tenure and rural development*. Food and Agricultural Organization.

Gaveau, D. L. A., Sheil, D., Husnayaen, S., Salim, M. A., Arjasakusuma, S., Ancrenaz, M., Pacheco, P., & Meijaard, E. (2016). Rapid conversions and avoided deforestation: Examining four decades of industrial plantation expansion in Borneo. *Scientific Reports, 6*, 32017. https://doi.org/10.1038/srep32017

Huijnen, V., Wooster, M. J., Kaiser, J. W., Gaveau, D. L. A., Flemming, J., & Parrington, M. (2016). Fire carbon emissions over maritime southeast Asia in 2015 largest since 1997. *Scientific Reports, 6*(26886), 1–8. https://doi.org/10.1038/srep26886

Rights and Resources Initiative. (2016). *Closing the gap: Strategies and scale needed to secure rights and save forests*. Rights and Resources.

Amnesty International. (2018). *Harvest of death: Three years of bloody clashes between farmers and herders in Nigeria*. Amnesty International.

Koplitz, S. N., Mickley, L. J., Marlier, M. E., Buonocore, J. J., Kim, P. S., Liu, T., Sulprizio, M. P., DeFries, R. S., Jacob, D. J., Schwartz, J., Pongsiri, M., & Myers, S. S. (2016). Public health impacts of the severe haze in Equatorial Asia in September–October 2015: Demonstration of a new framework for informing fire management strategies to reduce downwind smoke exposure. *Environmental Research Letters, 11*(9), 1–10. https://doi.org/10.1088/1748-9326/11/9/094023

Lustgarten, A. (2018). Palm oil was supposed to help save the planet. Instead it unleashed a catastrophe. *New York Times*. https://www.nytimes.com/2018/11/20/magazine/palm-oil-borneo-climate-catastrophe.html

Masuda, Y. J., Kelly, A. C., Robinson, B. E., Holland, M. B., Bedford Malcolm, C., Game Chloe, C. E. T., Ginsburg Thea, Hilhorst Steven W., Lawry Daniela A., Miteva Jessica, Musengezi Lisa, Naughton-Treves William D., Sunderlin Peter, Veit (2020) How do practitioners characterize land tenure security?. *Conservation Science and Practice 2*(5). https://doi.org/10.1016/10.1111/csp2.186

McDonnell, T. (2018a). Herders vs. farmers: A deadly year in Nigeria. *National Public Radio*.

McDonnell, T. (2018b). Why it's now a crime to let cattle graze freely in 2 Nigerian states. *National Public Radio2*. https://www.npr.org/sections/goatsandsoda/2017/12/12/569913821/why-its-now-a-crime-to-let-cattle-graze-freely-in-2-nigerian-states

North, D. C. (1990). Institutions, economic theory, and economic performance. In *Institutions, institutional change and economic performance*. Cambridge University Press. https://doi.org/10.1017/CBO9780511808678.014

North, D. C. (1991). Institutions. *The Journal of Economic Perspectives, 5*(1), 97–112. https://doi.org/10.2307/1942704

Page, S. E., Siegert, F., Rieley, J. O., Boehm, H.-D. V., Jaya, A., & Limin, S. (2002). The amount of carbon released from peat and forest fires in Indonesia during 1997. *Nature, 420*, 61–65.

Rist, L., Feintrenie, L., & Levang, P. (2010). The livelihood impacts of oil palm: Smallholders in Indonesia. *Biodiversity and Conservation, 19*(4), 1009–1024.

Robinson, B. E., Masuda, Y. J., Kelly, A., Holland, M. B., Bedford, C., Childress, M., et al. (2018). Incorporating land tenure security into conservation. *Conservation Letters, 11*, e12383. https://doi.org/10.1111/conl.12383

Roos, M. C., Kessler, P. J. A., Robbert Gradstein, S., & Baas, P. (2004). Species diversity and endemism of five major Malesian islands: Diversity–area relationships. *Journal of Biogeography, 31*, 1893–1908.

Schlager, E., & Ostrom, E. (1992). Property-rights regimes and natural resources: A conceptual analysis. *Land Economics, 68*(3), 249–262. https://doi.org/10.2307/3146375

Sheil, D., Casson, A., Meijaard, E., van Noordwijk, M., Gaskell, J., Sunderland-Groves, J., et al. (2009). *The impacts and opportunities of oil palm in Southeast Asia: What do we know and what do we need to know?* CIFOR Occasional Paper.

Sjaastad, E., & Bromley, D. W. (2000). The prejudices of property rights: On individualism, specificity, and security in property regimes. *Development Policy Review, 18*(4), 365–389. https://doi.org/10.1111/1467-7679.00117

van Gelder, J. L. (2010). What tenure security? The case for a tripartite view. *Land Use Policy, 27*(2), 449–456. https://doi.org/10.1016/j.landusepol.2009.06.008

2

A Historical Perspective on Land Tenure Security

William D. Sunderlin and Margaret B. Holland

How Did We Get Here?

Why do vast numbers of people in rural areas of the Global South lack tenure security over the lands they use (RRI, 2015a; USAID, 2016)? We answer this question by examining how various processes in the course of socioeconomic development have produced this outcome. In doing so, we intend to show that the answer—far from being straightforward or obvious—is more complex than one might suppose. Our overarching argument is that understanding the past is essential for addressing present challenges tied to tenure insecurity.

Answering this question is encumbered by several obstacles. First, there are no historical measures of tenure security. Recall from Chap. 1

W. D. Sunderlin
Center for International Forestry Research (CIFOR), Bogor, Indonesia

M. B. Holland (✉)
Department of Geography & Environmental Systems, University of Maryland, Baltimore County, Baltimore, MD, USA
e-mail: mholland@umbc.edu

M. B. Holland et al. (eds.), *Land Tenure Security and Sustainable Development*,
https://doi.org/10.1007/978-3-030-81881-4_2

that we define tenure security as the landholder's perception that rights will be upheld by society (Sjaastad and Bromley, 2000). For lack of historical information on peoples' perceptions of land tenure security, we use information on past patterns of land dispossession and regaining of land control as a proxy for tenure insecurity and security. We draw attention to certain historical developments and trends that we believe have obvious (if not measured) impact on tenure security. Moreover, we look at episodes of contested land rights and their violations as a stand-in measure of security gained and lost. Parts two and three of this chapter, focusing as they do on a historical period when there is no measurement of tenure security, are necessarily conjectural and hypothetical. Parts four and five, which encompass the period after World War II, are more empirically grounded.

Second, we are undertaking a vast topic that could itself be a whole book, recognizing the relevance and importance of providing historical context for contemporary tenure security issues. Of necessity, our chapter is more a theoretical sketch with a few historical illustrations rather than the historical treatise that the topic deserves. We sacrifice detail to reveal the forces and trends that might otherwise escape our attention.

Third, the general propositions we make mask variations across and even within continents and countries. Because tenure security can vary by gender, ethnicity, and other social categories as well, a deeper analysis of tenure security requires careful study of the local context and history of any given population group of interest.

There are several assumptions and values underlying this effort that should be disclosed. We recognize that tenure security, in practice, has sometimes been a zero-sum situation where gains in tenure security of the powerful are often achieved at the expense of the security of the less powerful (Broegaard, 2005; Chomba et al., 2016; Robinson et al., 2018). With equity in mind, we are concerned mainly with the tenure security and rights of less powerful "common" people in rural areas. Recent reports by donor and development agencies imply the emergence of a normative focus on the tenure rights of the poor (IFAD, 2015; World Bank, 2019). Unfortunately, some donor initiatives push for large-scale impact through uniform approaches that fail to recognize the nuance required in attending to issues of equity in land rights and tenure security. We are concerned about those who have suffered in the development process, in particular, Indigenous Peoples and their customary systems, other traditional peoples, and peasant farmers (Gilbert, 2016; Komey, 2008; Lawry, 2014). We are concerned not just about land, but also about non-land resources above and below the land (e.g., water, forests, and minerals) because they too have been the object of appropriation in the development process. Moreover, we believe that customary land and resource claims have inherent value and can be legitimate even if not formally recognized by the state (Peters, 2009; Unruh, 2006). We believe hierarchy and class (e.g., who holds power) are key conceptual lenses for understanding the historically salient trends and shifts we document.

The chapter is comprised of five parts. In the next section we explain the ways socioeconomic development has often placed rural people in developing countries in a disadvantageous position with regard to their tenure security. In the third section we identify the structural and natural forces at work in the development process and how they produce varied outcomes. In the fourth section we look at major episodes in the ebb and flow of rights and tenure security. The concluding section points out the relevance of these historical insights as we enter an uncertain and volatile period.

Development and Dispossession

Across history, dramatic changes have occurred in the way humans have lived on planet Earth, due largely to the development and spread of capitalism. How did rural people view the security of their access to and control of land in the feudal centuries (mainly ninth through the fifteenth), during early capitalism (beginning in the seventeenth century) and at the time of the emergence of industrialism (beginning in the eighteenth century)? It's difficult to know with certainty for lack of information, but we can suppose it exhibited the wide range of possibilities we see today. At one extreme, peasants on *early* feudal estates (Editors of Encyclopedia Britannica, 2012) and slaves on eighteenth-century American plantations had no meaningful control over their livelihoods and had no land tenure security. At the other extreme, some customary hunting and gathering societies living far from cities and villages (and therefore far from the ravages of emerging capitalism) probably had reasonably strong confidence in future land and resource control—if we set aside the possibility of territorial conflict.

In what follows, we make the case that precarious access to and control over land would undergo changes in its causes, geographical scope, and intensity in the ensuing centuries. As socioeconomic change unfolded, the causes of land tenure insecurity were no longer just nation states, city states, kingdoms, and chieftains fulfilling territorial ambitions, but were also driven by centers of urban economic production seeking land (Royston, 2002; Unruh, 2007), increasing demand for natural resources (raw materials) and labor, and growing markets for selling their products (Cattaneo, 2001; Firmin-Sellers, 2000). The growing and shifting causes of land tenure insecurity first grew within and then across national and continental boundaries. Land markets and the process of formalization and commodification were factors that increased tenure security and insecurity (Deininger et al., 2011; Kelly & Peluso, 2015).

In medieval times under feudalism, and more specifically under the manorial system (organization of the economy under feudalism), aristocrats provided peasants (small farmers or landless laborers) military protection against outside aggression in exchange for services on land they supplied (La Croix, 2002; Ellsworth, 2002; Editors of Encyclopedia Britannica, 2012). These services were typically of four kinds: money; labor through the use of the peasants' own plow and oxen; reaping and processing of the harvest; and military service (Editors of Encyclopedia Britannica, 2012; Ellsworth, 2002). As explained above, early on, feudal tenure was insecure inasmuch as the lord could evict the peasant tenant at any time. With the passage of time feudal tenure became more secure. Through cultural change and recourse to royal courts, lords could no longer arbitrarily force peasants off the land, but instead had to ensure permanent access to that land (Editors of Encyclopedia Britannica, 2012). This arrangement signaled the beginning of the idea of secure land tenure (Bruce, 1998; Ellsworth, 2002). Feudalism went into decline from the eleventh century onward. With the growth of the money economy and of cities and towns, and with increasing demand for agricultural surplus, it became more efficient and profitable to have free workers who paid rent or received wages (Editors of Encyclopedia Britannica, 2012).

We define capitalism as "an economic system based on market competition and the pursuit of profit, in which the means of production or capital are privately owned by individuals or corporations" (OESD, 2020). While according to some observers, capitalism and globalization have played a role in reducing global poverty (Chandy & Gertz, 2011; Ortiz-Ospina, 2017), these same forces have triggered or aggravated uneven access to land, often to the detriment of rural tenure security in the Global South (UN-Habitat, 2014; UN, 2020). Exponential growth of capitalist production has translated to considerable geographic reach in the search for profits through additional land, resources, labor, and markets. Contemporary cases of this are discussed in the next section of this volume (Chap. 7) concerning large-scale land acquisitions (i.e., "land grabs") in parts of Africa. An important associated phenomenon, made possible in part through capitalist development (Magdoff & Foster, 2013), is the exponential net increase in the global human population—growing from approximately 650 million in 1750 to 7.7 billion now (a more than ten-fold increase). Just as important, from the standpoint of increasing resource pressure and extraction, is exponential average per capita growth in resource and energy consumption (although with considerable disparity) (FOE, 2009; Ritchie, 2020). The creation of a largely urban consumer class underlies the vitality of capitalist development and the growth in consumption (Naik & Oldfield, 2015; Ghosh, 2019).

No less important in terms of rural impact have been the ways in which early capitalism actively suppressed land rights. This was not limited to the Global South. In England, for example, the Enclosure Acts of 1750–1850 overtook the common lands used by small farmers, preventing their access and creating the basis for much larger farms owned by the wealthy (Rosenman, 2012). Many rural people who had heretofore relied on those lands for survival were forced to migrate to urban areas in search of wage labor opportunities.

International commodities trade long predates the birth of capitalism and the industrial revolution. But the onset of capitalism and industrialization led to *colonialism* (the practice of acquiring control over another country, occupying it with settlers, and exploiting it economically) and *imperialism* (the practice of extending a country's power and influence through diplomacy or military force), which shifted the search for land, labor, and markets from the industrializing North to Africa, Asia, and Latin America.

A combination of economic drivers, woven together with religious and racial ideologies, were key to motivating European white settlers to suppress the rights of other peoples and lay claim to vast areas of land around the world. The Doctrine of Discovery, proclaimed as a papal edict in the fifteenth century, laid the groundwork for the outlook that white Christians were intellectually and morally superior to non-white and non-Christian people and that "discovery" and occupation of lands inhabited by "barbarians," "savages," and "heathens" was an essential first step in the civilizing process (Mark & Soong-Chan, 2019; Miller et al., 2012; Special Rapporteur on Human Rights, 2010). The idea of racial superiority manifested itself in a wide array of proclamations and campaigns aiming at land seizure across countries of the Global South, as well as North America (Bonds & Inwood, 2016; Smith, 2012).

In the eighteenth and nineteenth centuries, opportunities for the production of cotton and tobacco in the United States and sugar in Brazil led to the importation of hundreds of thousands of slaves from Africa to meet growing agricultural production goals (Galeano, 1997). The search for gold in the Andes, mining ores in the Congo Basin, and spices in the Netherlands Indies were the entry point for incipient colonization in those regions. There were genocidal conflicts that greatly reduced Indigenous populations and constrained them to restricted areas in large areas of the South. European settler populations subjugated wide areas of the globe, created governance structures that turned countries into dependent clients, and organized production and development activities to be of direct service to foreign occupying armies and economic elites (Braudel, 1979; Cardoso & Faletto, 1979; Wolf, 1982; Harvey, 1982).

The simple sketch we have given of this apparent "march of history" suggests that the process of socioeconomic development has led relentlessly and uniformly to the appropriation of lands and suppression of land and resource access in remote rural areas. But this is not the case. As we shall see in the next section, in which we decompose some of the basic elements of the development process, the trajectory of change in land tenure security is far more complex than a cursory glance would suggest.

Development Dynamics and Tenure Security Complexity

We identify four forces at work in the development process that help explain the variability and complexity of pressures on access to land and resources across time (what we call tenure security today): vertical social interactions; horizontal social interactions; technological dynamism; and catastrophes (both natural and human-induced). Some of the synergies among these four forces that will be alluded to are for illustrative purposes. Although it would have been possible to include a fifth category on tenure policies, we chose not to, seeing that such policies are in a sense subsumed in, and artifacts of, the other categories. Moreover, we recognize that, in the land tenure literature, attention to policies tends to overshadow the structural forces shaping them. Several chapters in the following sections of this volume explore different land tenure policies, including the post-colonial period of land reform and formalization (Chap. 11) and more recent national and global policy shifts toward recognition of Indigenous Peoples and local communities (Chap. 4).

Vertical Power Interactions

People lose access to land or can feel insecurity at the perceived risk of losing the land they have long held and used because someone, or some group, more powerful than them can take it away. Vertical land pressure (induced by hierarchies of power) is a constant across recorded history. The more powerful actors seeking land or resources could be an invading army, the state (acting either through the threat of violence or eminent domain), or a private enterprise. A key characteristic in these interactions is the use of violence, intimidation (threat of violence), or other threat of reprisal (e.g., legal confiscation or economic penalties) by the powerful actor(s) to accomplish their goals. Across time the threat of land appropriation has become more sophisticated, layered (e.g., collaborative and coordinated activities of the military, state, private enterprises, and groups of individuals), and far-reaching in a geographical sense. In the past, land appropriation often involved direct violence linked to increasingly sophisticated use of weaponry, counterinsurgency, rape, and torture as an instrument of conquest, and other forms of subjugation (e.g., Escobar, 2004; Daudelin, 2003; Grajales, 2013; Ward & Marsh, 2006). In recent decades there has been increased use of forms of subjugation that are less visibly brutal and achieve their goal through economic coercion, negotiation, and manipulation (e.g., Dell'Angelo et al., 2017; Regassa et al., 2018; Alden Wily, 2012). Also, in recent decades, there has been a growing role of land speculation, offshore land investment, loan sharking, drug trafficking, and coercive conservation in land appropriation (e.g., McMichael, 2012; McSweeney & Pearson, 2016; Peluso, 1993; Agrawal & Redford, 2009). This trend is further discussed in Chapter 7 of this volume.

Across history there has always been resistance to vertical oppression and threats of land appropriation. Although often futile (in the sense that land is often seized precisely because the claimant is more powerful), organized resistance movements have had some notable successes. Examples are agrarian movements involving guerrilla warfare against oligarchies in Latin America (Stavenhagen, 1970; Teubal & Ortega Breña, 2009), the deployment of "weapons of the weak" (covert but powerful resistance to domination through non-cooperation or cultural resistance) (Scott, 1998) to undermine elite control and act in defense of land and resources, and international and national coalitions in support of Indigenous land rights that have been helpful in achieving formal statutory Indigenous land claims (e.g., Anaya & Grossman, 2002).[1]

Horizontal Social Interactions

The sources of land pressure are also horizontal, in the sense of being created by the multiplied presence of people with more or less equal power. Exponential and accelerating rates of human population growth—with the highest rates experienced recently in the Global South—have greatly increased population density, land scarcity, and competition in rural areas. Note that there are important synergies with vertical pressure. For

[1] The breakthrough international court case of Awas Tingni in Nicaragua opened key legal pathways for claims by Indigenous Peoples all over the world.

example, in Latin America the seizure of vast areas of the best lands for as pasture by rich landlords led to a growing number of land-poor and landless peasants competing for remaining land or migrating to cities (Carlson, 2019; Kay, 1997; Shaw, 1974). Land reforms that may have been able to redistribute lands to the rural poor instead avoided doing so in favor of colonizing rainforest areas, which were lands often already customarily-held by Indigenous groups (Jones, 1990). Migration is a key factor in land scarcity and pressure (Bilsborrow, 1992; Bilsborrow & DeLargy, 1990; Carr, 2009). Migrations to settled areas can be the result of seasonality (e.g., for fruit harvesting and pastoralism); state-sponsored relocation (e.g., Indonesia's transmigration program of the 1980s to reduce land pressure in Java (Fearnside, 1997)); war involving religious, ethnic, or racial persecution (e.g., refugees in search of security on either a temporary or a permanent basis); or natural cataclysm (more on this further in the chapter).

Although there has been considerable human population growth and an increase in displacement and human migrations, this does not translate monotonically to increased rural land pressure. Some of the release in pressure is planned (e.g., Indonesia's transmigration), but much of it is spontaneous and unplanned. A key characteristic of capitalist development has been the growth of urban areas, where much of manufacturing and industrial production takes place and where there is a growing service and tertiary sector. The rural "push factor" (land and employment scarcity and low income) is matched by the urban "pull factor" (relatively higher wages). Net rural outmigration in developing countries is part of what explains the steady growth of urbanization in countries of the Global South (Keats & Wiggins, 2016:5). Economic crisis can reverse this process. Examples are the "return to the village" in Cameroon in the aftermath of the devaluation of the CFA franc in 1985 (Franqueville, 1987), and the current urban to rural migration resulting from COVID-19 (Shylendra, 2020; Boillat & Zähringer, 2020).

However, in net terms urbanization has proceeded across the planet, with Africa as the only remaining continent with majority rural population (UN, 2019). Another pressure relief valve is intensification in agricultural production. Greater yields on a unit of land through technological development (see below) can potentially contribute to relieving land scarcity and competition (Byerlee et al., 2014). The Danish economist Ester Boserup demonstrated that increased agricultural population density can induce innovation and increase production, contrary to Malthusian theory (Boserup, 1965).

Technological Dynamism

In hundreds of ways that cannot all be catalogued here, technological innovation is at the heart of the capitalist and industrial revolutions that have expanded the consumption of resources and appropriation of lands in rural areas. Invention of the steam engine in the seventeenth century, closely followed by the fossil energy-fueled internal combustion engine, greatly increased the capacity and efficiency to access natural resources, transform them into processed consumer goods, and deliver them to markets. The transportation revolution (from walking, to animal draught, sail boats, engine-powered boats, railroad system, and air travel) meant not only improved speed, capacity, and efficiency of transport, but also increase in the means of the military, state, and private enterprise to conduct their work at increasing distances. Likewise, the communications revolution (voice, written letter, telegraph, telephone, electronic mass media, Internet) contributed to the same outcome.

Yet in all sorts of identifiable ways, technological dynamism has also contributed to reducing land pressures in rural areas. Agricultural intensification through irrigation, fertilizers, pesticides, fungicides, and high-yielding varieties of seeds has not only increased yields on per unit areas of lands and increased the number of harvests in a year (Pingali, 2012), but the application of technology is viewed as key to future global food security (Rosegrant et al., 2014; Fuglie et al., 2020). Agricultural intensification, diversion of rural labor from agricultural to non-agricultural activities, and migration to urban areas can logically make rural land pressure lighter than it would otherwise be. As explained by García et al. (2020), agricultural intensification can spare land for nature (i.e., reduce agricultural land demand), but it can also have a rebound effect involving further expansion of cropland. Technological development is an essential facet of rural-to-urban migration described above. The relatively high subsistence wage in urban areas is directly related to increases in productive efficiency in the manufacturing and industrial sectors over time. Rural to urban migration is motivated not just by economic factors (higher average wages), but also by security (e.g., avoidance of war and conflagration) and culture. These two-edged effects of technology are also evident, for example, in the case of advancements in communication. Electronic dissemination of information (news, entertainment) to remote regions has drawn attention to alternative life experiences outside the rural realm and made migration to the city attractive to some. However, the same communication advances, notably the cell phone, have also enabled some agriculturalists to increase production and income, enhancing the security of rural livelihood (Aminou et al., 2018; Fabregas et al., 2019).

Catastrophes—Natural and Human-Induced

Since time immemorial natural disasters of various kinds have been a threat to rural people's access to and control over land. There is ample documentation across the centuries of mass migration provoked by natural disasters such as floods, droughts, fires, earthquakes, landslides, pestilence and disease (such as the plague), insect infestation and consequent crop decimation, earthquakes, and tidal waves. There are also catastrophes caused by human agency that can profoundly disturb rural land tenure security. Examples are excessive or inappropriate resource use and consequent drop in productivity (e.g., over-grazing, hillside agriculture, and landslides) and industrial disasters (e.g., Bhopal chemical disaster in India). The most notable contemporary case is the climate crisis, which is an exacerbating factor for droughts, extreme heat events, hurricanes, and other natural disasters. The World Bank has predicted that climate change will force 140 million people to migrate by 2050 (Rigaud et al., 2018). Increasing average temperatures are also altering the optimal locations for the production of certain crops. Producers of these crops will be forced to either adapt locally or relocate. Evidence is emerging that hot temperatures are decreasing labor hours in rural communities in tropical low-latitude countries, threatening existing livelihood activities that are often tied to the land (Masuda et al., 2019). Rising seas threaten inundation of coastal lands, which is where the highest densities of human populations currently live (Podesta, 2019; Le, 2020). It is forecast that the most devastating consequences of threatened crop production will happen in rural areas of developing countries (Morton, 2007; Mendelson, 2008; Nelson et al., 2009; FAO, 2016; FAO, 2017; Arnall, 2019; Sloat et al., 2020). Chapter 2c of this volume explores the relationship between this shifting and increasing food insecurity and land tenure insecurity.

Historical Trends in the Ebb and Flow of Security and Insecurity

We have seen that, in broad terms, the diversity, geographical reach, and intensity of threats to rural tenure security have tended to increase over time—with considerable variation across and within countries, and with forces that increase insecurity often encountering other forces that decrease it. Are there any clear net tendencies in the ebb and flow of tenure security across space and time? It is not possible to answer this question with a high degree of specificity because of insufficient information, especially from the distant past, and because of the wide diversity of patterns across countries. Nevertheless, we can illustrate trends by pointing to a sequence of deprivation, gains, and rollback in the last century.

Roughly in the first half of the twentieth century we can see increasing threats to rural land access across the developing world. In Asia and Africa, capitalist penetration into the countryside was implemented through strong colonial domination, often facilitated by a captive national government acting as a client and surrogate to foreign powers (Newbury, 2000). Resistance to seizure of land and resources, and to colonialism in general, was uneven among countries (Maddox, 1993; Tussing, 2017; Chandavarkar, 1998; Bogaerts & Raben, 2012). Cities were not yet absorbing a large flow of migration from the countryside, with a consequent build-up of rural population. In the agricultural sector, there was wholesale appropriation of lands by occupying powers and rural elites, and increasing rates of landlessness (Frankema, 2010). In Latin America the dynamics were similar but with a tendency toward direct rather than indirect rule. By the early 1900s, the extremely inequitable land ownership system (called "latifundio-minifundio")[2] had already taken hold throughout the region, as demonstrated by Gini coefficient data from that era: 78 in 1920 Brazil, 80.3 in 1914 Argentina, and 83.7 in 1927 Chile (Frankema, 2008).[3] The Mexican revolution (1910–1920) was the best organized, earliest, and most successful movement to fight and partially reverse this inequality (McLynn, 2002). It resulted in the legal recognition of the *ejido* in the 1917 Mexican Constitution, which set the stage for land redistribution and the formal recognition of communal agrarian landholding (Perramond, 2008).

Across all three regions of the Global South, national governments carried forward eighteenth- and nineteenth-century state ownership over forest estates; the aim was to assert control over resources deemed strategic and to occupy remote areas for the protection of national borders (Peluso, 1992; Scott, 1998; Fay & Michon, 2012). Appropriation of forests as part of the national estate caused tenure security and displacement for Indigenous Peoples in remote regions. Forest estates were then exploited for timber, minerals, and petroleum and subsequently converted to plantation agriculture—a pattern that intensified in the second half of the twentieth century. Wars of national liberation and/or guerrilla movements across the developing world in the middle of the twentieth century were in part a response to rural exploitation (Wolf, 1969).

By the second half of the twentieth century, most developing countries gained independence from colonial powers. Nevertheless, post-colonial elites in many countries deepened control over land and resources and exploitative practices. Under neo-colonialism, national elites (in government and in the private sector) continued to benefit financially through trade and aid relationships with wealthier countries and consolidated a process of national capital accumulation for their own gain. During this period, leftist movements acquired national control and waged war

[2] "This dualistic tenure system is characterized by relatively few large commercial estates known as latifundios, which are over 500 hectares and numerous small properties known as minifundios, which are under 5 hectares. Minifundios are mainly subsistence-oriented smallholdings and are generally farmed by peasant households" (Wikipedia, 2020a).

[3] The Gini coefficient for land ranges from perfect equality at 0 to perfect inequality at 100. For comparison, the earliest data from Indonesia (1963) show a Gini of 52.7 (Frankema, 2008).

against former colonial powers (e.g., Congo and Vietnam). In the 1960s and 1970s we witnessed the emergence of leftist social movements in defense of land rights, and in some countries, guerrilla warfare was waged against national governments allied with oligarchies, for example, in El Salvador (Wood, 2003) and Nicaragua (Baracco, 2005). Movements for land tenure reform and land redistribution aimed to right extreme inequalities in the agricultural sector. In the Americas, agrarian reforms were undertaken in almost every country, sometimes by left-wing governments (e.g., Nicaragua and Cuba), but also by moderate governments responding to external and internal pressures and attempting to diffuse social unrest (de Janvry, 1981; Kay, 1997). Some analysts claim the transformations made by those reforms are disappointing: they were poorly or partially implemented, unleashed new conflicts, and were met with counter-reforms that, in the end, benefited capitalist farming while further marginalizing peasant farming (e.g., Kay, 1997). Lipton (2009) observes, "At least 1.5bn people today have some farmland as a result of land reform, and are less poor, or not poor, as a result. But huge, inefficient land inequalities remain, or have re-emerged, in many low-income countries. Land reform remains both 'unfinished business' … and alive and well" (Lipton, 2009: p. 8). A 2020 study finds that, in most countries, land inequality is increasing and that: "This trend directly threatens the livelihoods of an estimated 2.5 billion people worldwide involved in smallholder agriculture" (Anseeuw & Baldinelli, 2020:7).

More recently, national and sub-national movements in the forest sector emerged to defend forest land rights as well as to demand forest tenure reform (Larson & Dahal, 2012). Most notably in the 1970s and 1980s, some developing country governments began to devolve forest management to Indigenous People and local communities; the initial motivation of this devolution was forest restoration rather than the recognition of rights, but also with the goal of conserving and sustainably managing forests (Larson & Dahal, 2012). Various factors motivated this change, including acknowledgment that state-led forest management has been a failure; willingness of governments and the private sector to relax their grip on lands that have already been stripped of most of their timber wealth; a worldwide trend toward decentralized governance; greater acceptance of collective and customary systems; and increasing effectiveness of international and national campaigns to recognize the human rights and resource rights of Indigenous Peoples (Larson & Dahal, 2012; Barry et al., 2010). The area devolved is documented in detail by the Rights and Resources Initiative (RRI) and updated every five years (RRI, 2020). As explained by RRI, "As of 2017, Indigenous Peoples, Afro-descendants, and local communities had legally recognized rights to 15.3 percent of the world's forests, a 40 percent increase from 2002. Over 98 percent of this progress occurred in developing countries. Communities now have legal rights to 28 percent of the developing world's forests in Africa, Asia, and Latin America" (RRI, 2020).

This notable progress notwithstanding, in the last 10–15 years, in some developing countries, we are witnessing a slowdown and the threat of rollback of some gains for land rights and tenure security made in the last half century. Dating from the time of the 2007–2008 global recession, agribusinesses, governments, and speculative investors in wealthier countries have appropriated 20–60 million hectares of land (about 1% of agricultural or 1% of forest lands worldwide) (Wikipedia, 2020b). Through this "land-grabbing" phenomenon—defined as very large land acquisition through either buying or leasing—investors have aimed at food production (about 37% of investments) and biofuel production (about 20% of investments). Among the motivations have been the emergence of shortages of arable lands in richer or rapidly growing economies, and the aim to produce a larger share of food, fiber, and fuel abroad and import them. Another factor is recovery from the effects of the global economic recession, which temporarily lessened rural investment; it has since been restored. A disproportionate share of this activity has been in Asia, notably Indonesia, Malaysia, and India (Von Braun & Meinzen-Dick, 2009; Borras Jr et al., 2011; GRAIN, 2016; Land Matrix, 2020; Wikipedia, 2020b). Land grabbing has also involved appropriation of water resources (Rulli et al., 2013). In terms of the typology explained earlier, this phenomenon is best understood in the sphere of vertical dynamics and can be viewed as a reassertion of coercive power by elites at the top of international and national power hierarchies. (For more discussion of land grabbing as a more recent phenomenon, see chapter 2d.)

In the forest sector in some countries, there has been a slowing or even a reversal of progress on extending tenure rights to Indigenous Peoples and communities (RRI, 2015b pages 19–21; RRI, 2018 pages 21–23). This includes some areas appropriated in the interest of conservation (Agrawal & Redford, 2009). In recent years, there has been a growing number of murders, death threats, acts of sexual violence, and legal and illegal intimidation against people in resource-rich areas in developing countries (Sunderlin et al., 2018). In 2016, there were at least 201 forest defenders murdered, followed by 197 victims in 2017, in various conflicts over land and resources (Global Witness, 2017). A 2017 letter composed by rights defenders in 29 countries demanded that the United Nations urge governments to increase legal protection from violence. The letter states: "We need global action to counter the threats we face. This is not just a struggle for resources, it's a struggle for justice and social equality" (Human Rights Defenders, 2017).

The Current Situation

In countries of the Global South, tenure insecurity is widespread for common rural people in both agricultural and forest landscapes. Knowing how we got into this situation is vital because it gives insights into the scope of opportunities for deflecting further threats, maintaining gains achieved, and improving tenure security.

The variability of tenure security is firmly rooted in the dynamics of capitalist development. We therefore need to be attentive to the contemporary permutations of this economic system which so pervasively affects human existence and natural resource use on planet Earth. Attention to capitalism is often set aside as a given or ignored as a topic too intractable, large, or disturbing to talk about. Its importance is in direct proportion to its absence from the discussion.

With this theoretical framework as a backdrop, we have seen that there are four overarching factors at play in historical capitalist development that can both aggravate or ameliorate rural tenure security in the Global South: vertical power interactions; horizontal social interactions; technological dynamism; and catastrophes whether natural or human-made. There are cross-synergies among these four factors that are alluded to notionally in this chapter that deserve further research attention.

Furthermore, we have postulated a succession of epochs in rural tenure security: undermining of tenure security under rapid expansion of rural capitalist development and colonial domination in the early twentieth century; some (by no means thoroughgoing) improvement of tenure security through rural to urban migration, resistance to oppression, and devolution of tenure rights to some Indigenous Peoples and local communities in the late twentieth century; and undermining of tenure gains in recent years in relation to the land-grabbing phenomenon, rollback of some rights, and human rights violations against defenders of land and resource rights. We emphasize that there is a great deal of variation in these tendencies across and within countries.

We are entering an uncertain and potentially volatile period for tenure security. On the one hand, rhetorical commitment to the importance of secure tenure rights has never been stronger among multilateral agencies, international development organizations, and the donor community, as evidenced in the current set of Sustainable Development Goals (SDGs) and related targets. Notably, there has been official recognition of the importance of Indigenous management of tropical forests and associated strengthening of tenure rights in relation to global climate change mitigation (IPCC, 2018). On the other hand, land grabbing, rollback of rights, and violence against land and resource rights activists have all increased.

In this context, we believe it is appropriate to call for an ever-stronger commitment to upholding and strengthening tenure security for rural people by governments, donor organizations, and NGOs. Relatedly, it is important to expand research on tenure security to propel and support this stronger commitment.

Acknowledgment We wish to acknowledge the many valuable criticisms and suggestions made by Anne Larson and Nancy Peluso on an early draft of this chapter. We take full responsibility for any errors or omissions that might remain.

References

Agrawal, A., & Redford, K. (2009). Conservation and Displacement: An Overview. *Conservation and Society, 7*(1), 1–10.

Alden Wily, L. (2012). Looking Back to See Forward: The Legal Niceties of Land Theft in Land Rushes. *Journal of Peasant Studies, 39*(3-4), 751–775.
Aminou, F. A. A., Houensou, D. A., & Hekponhoue, S. (2018). Effects of Mobile Phone Ownership on Agricultural Productivity in Benin: The Case of Maize Farmers. *Journal of Economics and Development Studies, 6*(4), 77–88.
Anaya, S. J., & Grossman, C. (2002). *The Case of Awas Tingni v.* A Step in the International Law of Indigenous Peoples. Arizona Journal of International and Comparative Law I.
Anseeuw, W., & Baldinelli, G. M. (2020). *Uneven Ground: Land inequality at the heart of unequal societies.* International Land Coalition. https://d3o3cb4w253x5q.cloudfront.net/media/documents/2020_11_land_inequality_synthesis_report_uneven_ground_final_en_spread_low_res_2.pdf
Arnall, A. (2019). Resettlement as Climate Change Adaptation: What Can Be Learned from State-Led Relocation in Africa and Asia? *Climate and Development, 11*(3), 253–263.
Baracco, L. (2005). Nicaragua; The Imagining of a Nation. Algora Publishing.
Barry, D., Larson, A., & Colfer, C. (2010). Forest tenure reform: An orphan of many uncles. In A. Larson, D. Barry, G. R. Dahal, & C. Colfer (Eds.), *Forests for people: Community rights and forest tenure reform* (pp. 19–41). Earthscan.
Bilsborrow, R. E. (1992). Population Growth, Internal Migration, and Environmental Degradation of Rural Areas in Developing Countries. *Eur. J. Popul, 8*, 125–148.
Bilsborrow, R. E., & DeLargy, P. F. (1990). Land Use, Migration, and Natural Resource Deterioration: The Experience of Guatemala and the Sudan. *Popul. Dev. Rev, 16*, 125–147.
Bogaerts, E., & Raben, R. (2012). *Beyond Empire and Nation: The Decolonization of African and Asian societies, 1930s-1970s*. Brill.
Boillat, S., & Zähringer, J. 2020. COVID-19, reverse migration, and the impact on land systems. Global Land Programme. https://glp.earth/news-events/blog/covid-19-reverse-migration-and-impact-land-systems
Bonds, A., & Inwood, J. (2016). Beyond white privilege: Geographies of white supremacy and settler colonialism. *Prog. Hum. Geogr, 40*, 1.
Borras Jr, S. M., Hall, R., Scoones, I., White, B., & Wolford, W. (2011). Towards a better understanding of global land grabbing: an editorial introduction. *Journal of Peasant Studies, 38*(2), 209–216. https://doi.org/10.1080/03066150.2011.559005
Boserup, E. (1965). *The Conditions of Agricultural growth: The Economics of Agrarian Change Under Population Pressure*. G. Allen & Unwin.

Braudel, F. (1979). *The Wheels of Commerce. Civilization & Capitalism 15th-18th Century, Volume 2*. Harper & Row.

Broegaard, R. J. (2005). Land Tenure Insecurity and Inequality in Nicaragua. *Dev. Change, 36*, 845–864. https://doi.org/10.1111/j.0012-155X.2005.00438.x

Bruce, J.W. (1998). Review of tenure terminology. Tenure Brief 8. Madison, Wisconsin: Land Tenure Center, University of Wisconsin. https://minds.wisconsin.edu/bitstream/handle/1793/22013/73_tb1.pdf?sequence=1&isAllowed=y

Byerlee, D., Stevenson, J., & Villoria, N. (2014). Does intensification slow crop land expansion or encourage deforestation? *Glob. Food Secur, 3*, 92–98.

Cardoso, F. H., & Faletto, E. (1979). *Dependency and Development in Latin America*. University of California Press.

Carlson, C. (2019). Agrarian Structure and Under-Development in Latin America: Bringing the Latifundio "Back In.". *Lat. Am. Res. Rev, 54*, 678–693.

Carr, D. (2009). Rural Migration: The driving force behind deforestation at the settlement frontier. *Prog. Hum. Geogr, 33*, 355–378.

Cattaneo, A. (2001). Deforestation in the Brazilian Amazon: Comparing the Impacts of Macroeconomic Shocks, Land Tenure, and Technological Change. *Land Econ, 77*, 219. https://doi.org/10.2307/3147091

Chandavarkar, R. (1998). *Imperial Power and Popular Politics: Class, Resistance and the State in India, 1850-1950*. Cambridge University Press.

Chandy, L., & Gertz, G. (2011). Globalization Reduced Poverty. YaleGlobal Online. https://yaleglobal.yale.edu/content/globalization-reduced-poverty

Chomba, S., Kariuki, J., Lund, J. F., & Sinclair, F. (2016). Roots of inequity: How the implementation of REDD+ reinforces past injustices. *Land Use Policy, 50*, 202–213. https://doi.org/10.1016/j.landusepol.2015.09.021

Daudelin, J. (2003). *Land and Violence in Post-conflict Situations*. The North-South Institute.

Deininger, K., Ali, D. A., & Alemu, T. (2011). Impacts of Land Certification on Tenure Security, Investment, and Land Market Participation: Evidence from Ethiopia. *Land Econ, 87*, 312–334. https://doi.org/10.3368/le.87.2.312

Dell'Angelo, J., D'Odorico, P., Rulli, M. C., & Marchand, P. (2017). The Tragedy of the Grabbed Commons: Coercion and Dispossession in the Global Land Rush. *World Development, 92*, 1–12.

Editors of Encyclopaedia Britannica. (2012). Feudal land tenure [WWW Document]. Encycl. Br. https://www.britannica.com/topic/feudal-land-tenure

Ellsworth, Lynn. (2002). A Place in the World: Tenure Security and Community Livelihoods. Washington, DC: Forest Trends. https://www.forest-trends.org/wp-content/uploads/imported/Place_World.pdf.pdf

Escobar, A. (2004). Development, Violence, and the New Imperial Order. *Development, 47*(1), 15–21.

Fabregas, R., Kremer, M., & Schilbach, F. (2019). Realizing the Potential of Digital Development: The Case of Agricultural Advice. *Science, 366*(6471), 3038. https://doi.org/10.1126/science.aay3038

FAO. (2016). *The State of Food and Agriculture: Climate Change, Agriculture, and Food Security*. Food and Agriculture Organization of the United Nations.

FAO. (2017). *Migration, Agriculture, and Climate Change: Reducing Vulnerabilities and Enhancing Resilience*. Food and Agriculture Organization of the United Nations.

Fay, C., & Michon, G. (2012). Redressing Forestry Hegemony When a Forestry Regulatory Framework Is Best Replaced by an Agrarian One. *Forests, Trees, and Livelihoods, 15*(2), 193–209.

Fearnside, P. (1997). Transmigration in Indonesia: Lessons from its environmental and social impacts. *Environ. Manage, 21*, 553–570.

Firmin-Sellers, K. (2000). Custom, Capitalism, and the State: The Origins of Insecure Land Tenure in West Africa. *J. Institutional Theor. Econ, 156*, 513–530.

FOE. (2009). Overconsumption? Our Use of the World's Natural Resources. Friends of the Earth. https://cdn.friendsoftheearth.uk/sites/default/files/downloads/overconsumption.pdf

Frankema, E. (2008). *The Historical Evolution of Inequality in Latin America A comparative analysis, 1870-2000*. Publisher.

Frankema, E. (2010). The colonial roots of land inequality: geography, factor endowments, or institutions? *Economic History Review, 63*(2), 418–451.

Franqueville, A. (1987). *Une Afrique entre le village et la ville: Les migrations dans le sud du Cameroun*. Editions de l'ORSTOM.

Fuglie, K., Gautam, M., Goyal, A., & Maloney, W. F. (2020). *Harvesting Prosperity: Technology and Productivity Growth in Agriculture*. The World Bank.

Galeano, E. (1997). *Open veins of Latin America: Five centuries of the pillage of a continent*. NYU Press.

García, V. R., Gaspart, F., Kastner, T., & Meyfroidt, P. (2020). Agricultural Intensification and Land Use Change: assessing country-level induced intensification, land sparing and rebound effect. *Environmental Research Letters, 15*, 085007. https://doi.org/10.1088/1748-9326/ab8b14

Ghosh, I. 2019. Mapped: The Dramatic Global Rise of Urbanization (1950–2020). Visual Capitalist. Retrieved October 22, 2020, from https://www.visualcapitalist.com/map-global-rise-of-urbanization/

Gilbert, J. (2016). Land grabbing, investors, and indigenous peoples: New legal strategies for an old practice? *Community Dev. J, 51*, 350–366. https://doi.org/10.1093/cdj/bsv025

Global Witness. (2017). Defenders of the Earth. Global Witness.

GRAIN. (2016). *The global farmland grab in 2016: how big, how bad?* GRAIN. https://www.grain.org/article/entries/5492-the-global-farmland-grab-in-2016-how-big-how-bad

Grajales, J. (2013). State Involvement, Land Grabbing and Counter-insurgency in Colombia. *Development and Change, 44*(2), 211–232.

Harvey, D. (1982). *The Limits of Capital*. Verso.

Human Rights Defenders. (2017). Joint statement from Environmental rights defenders workshop, Johannesburg, 7-9 August 2017. 12 September 2017. http://www.civicus.org/index.php/media-resources/media-releases/open-letters/2939-joint-statement-from-environmental-rights-defendersworkshop

IFAD. (2015). Land tenure security and poverty reduction. International Fund for Agricultural Development, Rome, Italy.

IPCC. (2018). Climate Change and Land. Inter-governmental Panel on Climate Change. https://www.ipcc.ch/srccl/

de Janvry, A. (1981). *The agrarian question and reformism in Latin America*. Johns Hopkins University Press.

Jones, J. (1990). *Colonization and Environment: Land Settlement Projects in Central America*. UN University Press.

Kay, C. (1997). Latin America's Exclusionary Rural Development in a Neo-Liberal World. Presented at the 1997 meeting of the Latin American Studies Association, Guadalajara, Mexico.

Keats, S., & Wiggins, S. (2016). *Population change in the rural developing world: Making the transition. Briefing paper*. Overseas Development Institute. https://www.odi.org/sites/odi.org.uk/files/resource-documents/10570.pdf

Kelly, A. B., & Peluso, N. L. (2015). Frontiers of Commodification: State Lands and Their Formalization. *Soc. Nat. Resour, 28*, 473–495. https://doi.org/10.1080/08941920.2015.1014602

Komey, G. K. (2008). The denied land rights of the indigenous peoples and their endangered livelihood and survival: the case of the Nuba of the Sudan. *Ethn. Racial Stud, 31*, 991–1008. https://doi.org/10.1080/01419870701568940

La Croix, S. J. (2002). Land Tenure: An Introduction. Working Paper No. 02-13. Honolulu, Hawaii: East-West Center. file:///Users/William/Downloads/WP_02-13%20(1).pdf

Land Matrix. (2020). Land Matrix public database on land deals. Retrieved October 22, 2020, from https://landmatrix.org/#:~:text=The%20Land%20Matrix%20is%20an,%2C%20regional%2C%20and%20national%20level.

Larson, A. M., & Dahal, G. R. (2012). Forest tenure reform: New resource rights for forest-based communities? *Conservation and Society, 10*, 77–90.

Lawry, S. (2014). Customary Land Tenure. Dev. Altern. Retrieved March 16, 2020, from https://dai-global-developments.com/articles/customary-land-tenure.

Le, T. D. N. (2020). Climate change adaptation in coastal cities of developing countries: characterizing types of vulnerability and adaptation options. *Mitig Adapt Strateg Glob Change, 25*, 739–761. https://doi.org/10.1007/s11027-019-09888-z

Maddox, G. (1993). *Conquest and Resistance to Colonialism in Africa*. Library Editions; World Empires.

Magdoff, F., & Foster, J. B. (2013). *What Every Environmentalist Needs To Know About Capitalism*. Monthly Review Press.

Mark, C., & Soong-Chan, R. (2019). *Unsettling Truths: The Ongoing, Dehumanizing Legacy of the Doctrine of Discover*. InterVarsity Press.

Masuda, Y. J., Castro, B., Aggraeni, I., Wolff, N. H., Game, E., Garg, T., Krenz, J., Ebi, K., & Spector, J. (2019). How are healthy, working populations affected by increasing temperatures in the tropics? Implications for climate change adaptation policies. *Global Environmental Change, 56*, 29–40.

McLynn, F. (2002). *Villa and Zapata: A History of the Mexican Revolution*. Carroll and Graf Publishers.

McMichael, P. (2012). The Land Grab and Corporate Food Restructuring. *Journal of Peasant Studies, 39*(3-4), 681–701.

McSweeney, K., & Pearson, Z. (2016). Prying Native People from Native Lands: Narco Business in Honduras. *NACLA Report on the Americas, 46*(4), 7–12.

Mendelson, R. (2008). The Impact of Climate Change on Agriculture in Developing Countries. *Journal of Natural Resources Policy Research, 1*(1), 5–19.

Miller, R., Ruru, J., Behrendt, L., & Lindberg, T. (2012). *Discovering Indigenous Lands: The Doctrine of Discovery in the English Colonies*. Oxford University Press.

Morton, J. F. (2007). The Impact of Climate Change on Smallholder Subsistence Agriculture. *Proceedings of the National Academy of Science, 104*(50), 19680–19685.

Naik, D., & Oldfield, T. (2015). Urbanisation is not natural or inevitable. It's being inflicted upon us by the forces of capitalism. City Monitor.

Nelson, G. C., Rosegrant, M. W., Koo, J., Robertson, R., Sulser, T. Z., Ringler, C., Msangi, S., Palazzo, A., Batka, M., Magalhaes, M., Valmonte-Santos, R., Ewing, M., & Lee, D. (2009). *Climate Change: Impact on Agriculture and Costs of Adaptation*. International Food Policy Research Institute.

Newbury, C. (2000). Patrons, Clients, and Empire: The Subordination of Indigenous Hierarchies. *Asia and Africa Journal of World History, 11*(2), 227–263.

OESD. (2020). Entry on "capitalism." Open Education Sociology Dictionary. Retrieved October 22, 2020, from https://sociologydictionary.org/capitalism/

Ortiz-Ospina, E. (2017). Historical poverty reductions: more than a story about 'free-market capitalism'. Our World in Data. https://ourworldindata.org/historical-poverty-reductions-more-than-a-story-about-free-market-capitalism

Peluso, N. L. (1992). *Rich Forests, Poor People: Resource Control and Resistance in Java*. University of California Press.

Peluso, N. L. (1993). Coercing Conservation?: The Politics of State Resource Control. *Global Environmental Change, 3*(2), 199–217.

Perramond, E. P. (2008). The rise, fall, and reconfiguration of the Mexican ejido. *Geographical Review, 98*(3), 356–371.

Peters, P. E. (2009). Challenges in Land Tenure and Land Reform in Africa: Anthropological Contributions. *World Dev, 37*, 1317–1325. https://doi.org/10.1016/j.worlddev.2008.08.021

Pingali, P. L. (2012). Green Revolution: Impacts, Limits, and the Path Ahead. *Proceedings of the National Academy of Sciences, 109*(31), 12302–12308.

Podesta, J. (2019). *The Climate Crisis, Migration, and Refugees*. Brookings Institute.

Regassa, A., Hizekiel, Y., & Korf, B. (2018). 'Civilizing' the Pastoral Frontier: Land Grabbing, Dispossession, and Coercive Agrarian Development in Ethiopia. *Journal of Peasant Studies, 46*, 935–955.

Rigaud, K. K., de Sherbinin, A., Jones, B., Bergmann, J., Clement, V., Ober, K., Schewe, J., Adamo, S., McCusker, B., Heuser, S., & Midgley, A. (2018). *Groundswell: Preparing for Internal Climate Migration*. The World Bank.

Ritchie, H. 2020. Energy Production and Consumption. Our World in Data. Retrieved October 22, 2020, https://ourworldindata.org/energy-production-consumption

Robinson, B. E., Masuda, Y. J., Kelly, A., Holland, M. B., Bedford, C., Childress, M., Fletschner, D., Game, E. T., Ginsburg, C., Hilhorst, T., Lawry, S., Miteva, D. A., Musengezi, J., Naughton-Treves, L., Nolte, C., Sunderlin, W. D., & Veit, P. (2018). Incorporating Land Tenure Security into Conservation: Conservation and land tenure security. *Conserv.Lett, 11*, e12383. https://doi.org/10.1111/conl.12383

Rosegrant, M. W., Koo, A., Cenacchi, N., Ringler, C., Robertson, R. D., Fisher, M., Cox, C. M., Garret, K., Perez, N. D., & Sabbagh, P. (2014). *Food Security in a World of Natural Resource Scarcity: The Role of Agricultural Technologies.* International Food Policy Research Institute.

Rosenman, E. (2012). On Enclosure Acts and the Commons [WWW Document]. BRANCH Br. Represent. Ninet.-Century Hist. Retrieved February 17, 2020, from http://www.branchcollective.org/?ps_articles=ellen-rosenman-on-enclosure-acts-and-the-commons.

Royston, L. (2002). Security of Urban Tenure in South Africa: Overview of Policy and Practice, Chapter 10, in: Holding Their Ground: Secure Land Tenure for the Urban Poor in Developing Countries. Earthscan.

RRI. (2015a). *Who Owns the World's Land? A global baseline of formally recognize indigenous and community land rights*. Rights and Resources Initiative (RRI).

RRI. (2015b). *Looking for Leadership: New Inspiration and Momentum Amidst Crises.* Rights and Resources Initiative.

RRI. (2018). *At a Crossroads: Consequential Trends in Recognition of Community-Based Forest Tenure from 2012-2017.* Rights and Resources Initiative.

RRI. (2020). *Forest and Land Tenure*. Rights and Resources Initiative. Accessed October 21, 2020. https://rightsandresources.org/tenure-tracking/forest-and-land-tenure/

Rulli, M. C., Saviori, A., & D'Odorico, P. (2013). Global land and water grabbing. *Proceedings of the National Academy of Sciences, 110*(3), 892–897. https://doi.org/10.1073/pnas.1213163110

Scott, J. C. (1998). *Seeing like a state: How certain schemes to improve the human condition have failed.* Yale University Press.

Shaw, R. P. (1974). Land Tenure and the Rural Exodus in Latin America. *Econ. Dev. Cult. Change, 23*, 123–132.

Shylendra, H. S. (2020). How is Covid-19 playing out in rural India? The Hindu Business Line. May 25. https://www.thehindubusinessline.com/opinion/how-is-covid-19-playing-out-in-rural-india/article31671013.ece

Sloat, L. L., Davis, S. J., Gerber, J. S., et al. (2020). Climate adaptation by crop migration. *Nature Communications, 11*, 1243. https://doi.org/10.1038/s41467-020-15076-4

Smith, A. (2012). Indigeneity, Settler Colonialism, White Supremacy. In D. Martinez Hosang, O. LaBennett, & L. Pulido (Eds.), *Racial Formation in the Twenty-First Century*. University of California Press.

Special Rapporteur on Human Rights. 2010. Preliminary study of the impact on indigenous peoples of the international legal construct known as the Doctrine of Discovery. Submitted to Permanent Forum on Indigenous Issues. Ninth session.

Stavenhagen, R. (1970). *Agrarian Problems and Peasant Movements in Latin America*. Anchor Books.

Sunderlin, W. D., Larson, A. M., & Sarmiento Barletti, J.P. (2018). Land and carbon tenure: Some—but insufficient—progress. Chapter 8 in Arild Angelsen, Christopher Martius, Veronique De Sy, Amy E Duchelle, Anne M Larson and Pham Thu Thuy. 2018. Transforming REDD+: Lessons and new directions. Bogor, Indonesia: Center for International Forestry Research. pp. 93-102. http://www.cifor.org/publications/pdf_files/Books/BAngelsen1801.pdf

Teubal, M., & Ortega Breña, M. (2009). Agrarian Reform and Social Movements in the Age of Globalization: Latin America at the Dawn of the Twenty-first Century. *Lat. Am. Perspect, 36*, 9–20.

Tussing, N. 2017. African Resistance to European Colonial Aggression: An Assessment. Africana Studies Student Research Conference. 2. https://scholarworks.bgsu.edu/africana_studies_conf/2016/001/2

UN. (2019). *World Urbanization Prospects: The 2018 Revision*. United Nations, Department of Economic and Social Affairs, Population Division. https://population.un.org/wup/Publications/Files/WUP2018-Report.pdf

UN. (2020). World Social Report 2020: Inequality in a Rapidly Changing World. United Nations, Department of Economic and Social Affairs. https://www.un.org/development/desa/dspd/wp-content/uploads/sites/22/2020/02/World-Social-Report2020-FullReport.pdf

UN-Habitat. (2014). Land Tenure Security in Selected Countries: Synthesis Report. Nairobi, Kenya: United Nations Human Settlements Programme. https://unhabitat.org/sites/default/files/download-manager-files/Land%20Tenure%20Security%20in%20Selected%20Countries_English_2014.pdf

Unruh, J. D. (2006). Land Tenure and the "Evidence Landscape" in Developing Countries. *Ann. Assoc. Am. Geogr, 96*, 754–772. https://doi.org/10.1111/j.1467-8306.2006.00515.x

Unruh, J. D. (2007). Urbanization in the developing world and the acutely tenure insecure. *City, Debates, 11*, 115–120.

USAID, L.T. & R.M. (2016). *Why Land Rights Matter (Infographic)*. United States Agency for International Development.

Von Braun, J., & Meinzen-Dick, R. S. (2009). "Land grabbing" by foreign investors in developing countries: Risks and opportunities. IFPRI Policy Brief. Washington, DC: International Food Policy Research Institute.

Ward, J., & Marsh, M. (2006). Sexual Violence Against Women and Girls and its Aftermath: Realities, Responses, and Required Resources. A briefing paper prepared for Symposium on Sexual Violence in Conflict and Beyond. 21-23 June 2006. Brussels, Belgium.

Wikipedia. (2020a). Entry on Latifundio–minifundio land tenure structure. Retrieved October 22, 2020, from https://en.wikipedia.org/wiki/Latifundio%E2%80%93minifundio_land_tenure_structure#:~:text=The%20land%20tenure%20structure%20of,%E2%80%93minifundio'%20land%20tenure%20structure.&text=Minifundios%20are%20mainly%20subsistence%2Doriented,generally%20farmed%20by%20peasant%20households

Wikipedia. (2020b). Entry on "land grabbing." Retrieved October 22, 2020. "Land grabbing" in Wikipedia https://en.wikipedia.org/wiki/Land_grabbing

Wolf, E. R. (1969). Peasant Wars of the Twentieth Century. Harper & Row Publishers, 328 pp.

Wolf, E. R. (1982). *Europe and the People Without History*. University of California Press.

Wood, E. J. (2003). *Insurgent Collective Action and Civil War in El Salvador*. Cambridge University Press.

World Bank. (2019). Understanding Poverty: Land [WWW Document]. World Bank. Retrieved March 16, 2020, from https://www.worldbank.org/en/topic/land.

3

Who Defines Land Tenure Security? *De Jure* and *De Facto* Institutions

Brian E. Robinson and Moustapha Diop

Introduction

Let us begin with a simple thought experiment. Consider a space explorer heading off into the uncharted edges of the galaxy.[1] That explorer and their family find a new habitable planet with water, other life, and many other resources. Not too hot, not too cold. Luckily, they do not find other signs of any imminently dangerous or sentient life with whom to negotiate, so they decided to settle there in the peace and quiet as the first and only (human) family there. That early explorer family may choose to

[1] Loosely inspired by real-life possibilities envisioned in President Trump's last *Economic Report of the President* (USA) 2021 (Executive Office of the President Council of Economic Advisers, 2021).

B. E. Robinson (✉)
Department of Geography, McGill University, Montreal, QC, Canada
e-mail: brian.e.robinson@mcgill.ca

M. Diop
Department of Geography & Environmental Systems,
University of Maryland, Baltimore County, Baltimore, MD, USA

M. B. Holland et al. (eds.), *Land Tenure Security and Sustainable Development*,
https://doi.org/10.1007/978-3-030-81881-4_3

settle wherever they like. They end up at a floodplain that has plentiful fish in the river, freshwater, and nice sunset views. Grasslands nearby are easy to till and convert to some small agricultural plots (they brought seeds). They use the floodplain part of the year for farming and then retreat during the rainy season to avoid floods. There are also forests across the valley that house plenty of small wildlife creatures for hunting, new herbs to discover, various fungi, and woody species to use for building and heating. They use these multiple pieces of land throughout the year as they please. Without anyone else around, this single family has no other competition, no socially imposed limits or constraints on their activities or action. They can choose to use (or not use) resources as they please, only bound by environmental and biological constraints of the landscape and the climate. There are no other people with whom to negotiate any concept of fair use and, as such, there is no need to define the concept of who has a *right* to certain resources, to engage in activities, or otherwise draw benefits from the land.[2]

Eventually, others hear of the bounty that exists on this previously unknown planet and come to settle down. At first, some settle in other parts of the planet and face limited or no social pressures. But as time passes lands become more crowded, and some others come to settle in the same area as the first family. Wildlife is plentiful, and resources are abundant. As more newcomers arrive social relationships and agreements begin to develop. A community, in fact, is built based on common understandings of how land and resources will be used and managed by community members. Sometimes these understandings emerge through discussion, negotiation, and shared principles. Sometimes they emerge through conflict. With time, however, boundaries are defined, collectively agreed-upon (even if implicitly) rules of use are devised, and a functional and socially coherent society evolves.

The foundational basis of property rights is a social agreement. Members of a society must, even if implicitly, agree on who can and cannot use particular resources and in what way, and these agreements must be backed by a recognized authority (Bromley, 1992; Meinzen-Dick & Mwangi, 2009). In the space frontier example, as unoccupied areas are settled, rights emerge as understandings and agreements among community members. This is much how John Locke formalized ideas of the

---

[2] This is akin to John Locke's early ideas of a "state of nature" (Locke, 1689).

emergence of property rights in the seventeenth century (Locke, 1689) (see Chap. 2). Our understanding of property relations has developed since then (see Bromley, 2004), in part this is because it is exceedingly rare, if even possible, that there is "uncharted" territory in which other members of society do not have pre-existing claim or, at the minimum, have no interest (Banner, 2009; Ellis et al., 2021). Still, the example helps illustrate that as societies grow and land pressure increases, there becomes a need for clear and transparent processes that assign and enforce rights among various parties, and spell out the rules for how rights can be accessed, transferred, terminated, or gained. Locke even argued that the *primary* function of government is to secure and protect such property institutions (Locke, 1689). In any case, granting rights over land and property is like a social contract. As a society we agree to let some people have rights to some things and in turn agree that others do not have a right to that same thing.

This chapter briefly reviews how the "governors" of that social contract can exist (and throughout history, have existed) at various levels of society or administration. They can be part of a formal system of governance or equally a more informal system of social relations or local land management. The most common way land rights and land tenure is discussed is through contracts, titles, or deeds that are formalized by the state (here: government). However, the state is not always necessary or present for having rights—so-called informal rights can exist without a higher-level state sanction. These are often referred to as "customary" arrangements, or *de facto* community-level understandings of simply "how things work."

As we will discuss, legitimacy and security of rights can be strong in both formalized and *de facto* situations. The *security* of land tenure refers to how confident a landholder is that their land rights will be upheld by and within their community (Sjaastad & Bromley, 2000) (for key terms see Chap. 1). The community must be a recognized authority, but it is not always necessary that the authority is a centralized government (Diop, 1968; Pélissier, 2008). In some cases, especially when the state is absent or weak, household-level land tenure security may be more a function of local community dynamics and informal governance mechanisms. Still, in most current contexts, land pressures from outside parties are only increasing (e.g., see Chap. 7), meaning that sustained land tenure security must likely come with state-recognized backing of land rights. Here we review *de jure* (of the law) and *de facto* (common local practice) perspectives on where the authority over rights can be held, and their relationship to land tenure security.

Formal, State-Sponsored, *De Jure* Land Tenure

With the growth of the state and the increasing reach of capitalism (see Chap. 2), defining and adjudicating property rights over land and territory, even historically ignored remote areas, has become a common concern of national governments. When land rights and tenure over land are formalized, they are an institutionalized part of government and governance. State judicial and enforcement systems that adjudicate and uphold the rule of law also, similarly, uphold and enforce property relations.

Various types of rights can be conferred to individuals and groups which define the stream of benefits to which an entity is entitled from a parcel of land. A common way of describing types of rights is presented in Chap. 1 (see Table 1) as the right of access, withdrawal, management, alienation (the ability to sell), and due process of land (Schlager & Ostrom, 1992). Sets and subsets of these right "bundles" are often implied in common ways of talking about different types of property in terms of private, common, or public land (Robinson et al., 2018). For example, western thought traditions consider rights "well-defined" when all possible rights are held by a single entity or landholder, what is generally referred to as full ownership or private property (Cooter & Ulen, 2012). However, in some countries such as China, Ethiopia, Tanzania, and Mozambique, the state is the sole owner of the land and, as such, rights are never "well-defined" in a strict sense, since rights of alienation do not hold. What is still generally thought of as "private" land in these cases is land that is contracted from the state in long-term (leasehold) agreements, up to 70 years in the case of China (Ma et al., 2020; Wang et al., 2015). Still, when the substance (does the right exist?) and assurance (is the right upheld?) of rights (Sjaastad & Bromley, 2000) are clear to a household, leasehold agreements via long-term contracts can lead to high rates of land investment and productivity (Deininger et al., 2011; Lin, 1992), similar to expectations of freehold private property. With formalized private rights, governments typically develop parcel-based land registries, or a cadastral map, that document who has an "interest" in the land (see Chap. 11). When geolocated cadasters do not exist, land registries may hold legal recognition of land properties in the form of titles and deeds.

In cases where some rights or duties are shared between the public and private individuals, for example, in communal and public lands, rights may be formalized or *de jure* recognized by the state in various other ways. Some common terms that imply formalized rights include laws and regulations, zoning, policies, management plans, protected areas, or private easements. Box 3.1 describes these terms and how rights can be formalized within them.

Box 3.1 Common Terms Associated with Formalized (*De Jure*) Rights

Laws and regulations: Property rights or right bundles may be defined or outlined in specific laws or regulations enacted by the governing body.

Zoning: Federal, regional, or municipal zoning regulations, especially in more developed contexts, often define uses and restrictions on land use, which amount to rights enforcements handed down by the state. Zoning is also featured in public protected area management (in, e.g., biosphere reserves) by defining access, withdrawal, and sometimes management rights restrict certain activities to specific areas.

Policies: Often rights, duties, and responsibilities can change with specific and often shorter-term policies or programs that are put in place that may be more ephemeral than codified laws or regulations. Payment for ecosystem service programs and other incentive-based policy tools are effective contracts that landholders enter into with a governing body, that imply duties or restrictions in lieu of some compensation mechanism.

Management plans: National, regional, or local agencies can develop strategies or management plans with specific goals. These sometimes have land use restrictions or duties that must be upheld in adherence with what is determined to be in the public interest.

Protected areas: Protected areas are often developed to safeguard unique landscapes, ecosystems, and biodiversity in general. Overall goals of protected areas, or the subject of what is being "protected," can range from environmental to social, cultural, or historical (e.g., UNESCO World Heritage Sites). These areas have historically represented primarily top-down strategies that limit actions, behaviors, or uses of land. Also see note about protection in Zoning above.

Private easements: Sometimes referred to as fee-simple easements, these are legally codified restrictions on a privately owned land parcel. These effectively transfer some portion of landholders' rights to the state and have become a more common form of environmental protection for some developed regions, especially the USA and parts of Europe.

The categories in Box 3.1 constitute a simple descriptive (and far from an exhaustive) list. They are also not mutually exclusive of one another, as noted by the *Zoning* and *Protected area* categories, and can sometimes constitute overlapping or strategies that could be characterized in various ways. Similarly, laws, regulations, zoning, and policies are all somewhat overlapping, fuzzy, and sometimes synonymous categories. This list simply represents common terms that describe situations in which the state plays a key role in the formalization of land rights.

Additionally, it is important to recognize that formalization of land rights by the state does not equate to land tenure security for landholders (also discussed in Chap. 11). Too often security is equated with private land and formalized title, but simply creating rules or allocating rights through some of these formal mechanisms does not guarantee security of rights for an individual (van Gelder, 2010). Security requires that the proper institutions are in place to ensure that conflicts and claims can be adjudicated fairly and requires the landholder perceive those institutions are trustworthy and reliable. Governance needs to be clear and transparent, and the process by which individuals make a claim on their rights needs to be accessible.

A number of factors can limit the strength and efficacy of institutions and governance systems. For example, weak state governments may be understaffed or otherwise lack the capacity to monitor or enforce rights especially in more remote regions. This can create institutional imbroglio, where there may be a formal right on paper but little ability for a landholder to exercise that right in reality. For instance, in many countries in Sub-Saharan Africa, the decentralized local governance and elected officials in more rural communes do not have the necessary human and economic resources capabilities to properly manage and control individuals' tenure rights, at least in the way envisioned by the central government. In cases where formal governance is lacking, local governance fills in, which can be prone to be overly influenced by the locally powerful or allow for discrimination (Higgins et al., 2018). This "elite capture" manifests particularly in places where the land has high value and is attractive for domestic and international agri-business and extractives investors (Wolford et al., 2013). Therefore, when governance resources are limited, much of the monitoring and enforcement capabilities tend to be concentrated in population hubs and economic centers—the state may have a more limited capacity to develop its interest in more remote or far off areas (Bromley, 2008). As the old Chinese proverb says, "The mountains are high, and the emperor is far away."

Local, Self-Governed, and *De Facto* Land Tenure

The opening to this chapter described the imagined emergence of a local and self-governed property system. Of course, historically, local-level governance has been ubiquitous. Farming communities, Indigenous groups, nomadic pastoralists, and others have developed governing systems that worked for their communities. The growth of the nation-state has given rise to larger needs to formalize governance and rules, in ways that were not typically necessary beyond regionally agreed-upon systems. In many places where the state has not (yet) claimed an interest, customary land rights are the norm (see Chap. 4, which explores indigenous and customary tenure). This is especially true in Sub-Saharan Africa, where it is estimated around 78% of the land is under customary or "neocustomary"[3] control (Alden Wily, 2018). In these cases, the role of higher levels of governance and administration can be unclear, poorly enforced, or purposefully disengaged (via "devolution" of management or within autonomous regions, for instance). This leaves land tenure and land governance decisions to local communities.

Just as in the discussion above regarding statutory or formalized rights, *de facto*, or informal land tenure arrangements, can also span any combination of right bundles. What *de facto* cases have in common is that the local community (in some form), rather than a centralized government, is the source and arbiter of rights. These rights can be *socially* upheld and are often discussed in several ways. Some common ways *de facto* regimes are described include customary or traditional tenure, communal management, norms, and *de facto* management regimes. Box 3.2 describes these in more detail.

[3] Neocustomary tenure refers to cases where the state has attempted to recognize customary lands but generally allow for autonomous governance within that land (see Chimhowu, 2019).

Box 3.2 Common Terms Associated with Informal (*De Facto*) Rights

Customary or traditional tenure: Community-specified rights can be unique, highly locally specific, and often represent long-negotiated differences and nuances that defined the rights and duties for individuals versus the greater community. These are often referred to as "customary" tenure arrangements, generally representing a wide range of conditions that summarize heterogeneity in conditions rather than any kind of commonality between cases in this categorization.

Communal management: Similar to customary rights, reference to communal management systems is vague and lacks specificity needed to understand who holds which kinds of rights. Still, this term is often used to discuss a system where some kind of collective governance over land or resources is practiced, and generally with rules that are agreed upon and enforced by a community.

Norms: When individuals or members of society behave according to some kind or rule or practiced behavior that becomes normalized by the group, these are often referred to as norms. These behaviors maybe are internalized as part of "culture" or simply as "the way things are done" (e.g., inheritance norms that can vary widely across communities), but these can constitute sometimes unspoken rules about how land management and property relations are governed.

De facto *management regimes*: Land is sometimes managed without law or an overarching governance system. In such cases "*de facto* management" indicates that however land is being managed by local actors *is* its management system. Again, this can capture a huge range of local conditions.

Local and indigenous management systems have sometimes evolved over generations, responding to social and environmental stresses. Recent efforts to formalize some traditional and customary management systems aim to keep traditional, customary, or otherwise indigenous knowledge or norms intact while formally recognizing these informal institutions in the eyes of the state (Knight, 2010). Still, the adoption of customary institutions can have equity implications that do not align with modern sustainable development objectives, for example, when local rules reinforce overly local power imbalances or social, ethnic, or gender-based inequalities.

Importantly, individuals and households in informal land tenure arrangements can feel either secure *or* insecure that their *de facto* rights will be upheld by the community. Theories on the emergence of property rights as a social system note that private and individual rights are responses to competition and pressure for land and resources. Rights specify "how persons may be benefited or harmed and, therefore, who must pay whom to modify the actions taken by persons" (Demsetz, 1967: 347). But rules are costly to enforce, so it only makes sense to increase the strength of rights and protection of rights when a community feels the benefits conferred by the right are worth the cost of monitoring, enforcement, and adjudicating fair processes (Ostrom, 2002; Robinson et al., 2013).

Recent efforts show that in many cases perception of tenure security is often high, even without formalization of rights. For example, the Prindex project recently began collecting annual nationally representative data from several countries on the perception of rights. A recent report showed that while 83% of those with formal land documents felt secure in their rights, still 63% felt secure even *without* any formal documentation (Prindex, 2020). Thus, as discussed above, formalization is neither necessary nor sufficient to guarantee land tenure security (see Chap. 11 for further discussion). Local competition, gender, migration and population change, external land pressures from outside interests, and civil conflict can all affect the security of land rights (Ghebru & Lambrecht, 2017; Robinson et al., 2018; Stickler et al., 2017). Land tenure security in these informal and *de facto* situations often depends on how clear, transparent, and legitimate rights are within a community.

Informal Rights, Formalization, and Sustainability

Sources of insecurity exist for both formalized and informal land tenure arrangements. In some cases, the core source of tenure insecurity can be thought of as a lack of congruence between *de jure* and *de facto* rights. This "tenure gap" can be a source of conflict, confusion, and dispute that alone can manifest (Robinson et al., 2018). Closing the gap between these two is important and has been a core effort of many land tenure interventions in the past several decades (Tseng et al., 2021). This requires ensuring customary lands are recognized at the state level, and also that local rights are just and appropriate at the community level. Other challenges come when there are overlapping or conflictual claims, which can often be brought on by development pressures. Regional land disputes can be difficult to resolve, and indeed are the resource at the core of violence and war (see Chap. 8), as well as overlapping claims between local communities and the state, for example, indigenous lands versus protected areas (Holland et al., 2014). Still, addressing these may not be easy—managing and resolving land disputes can quickly become an exercise in conflict resolution and possibly involve generations of disagreements.

The formalization of rights is sometimes a precursor to development, and in other cases is a messy and conflict-ridden consequence of development (Fenske, 2011; Ho, 2015). Regardless, if the process of defining and determining rights begins when dispute or conflict is active, those rights are bound to be contested and fought over (see Chap. 8). Formal recognition of customary land rights is ideally proactive and happens prior to any imminent "need" for those rights to be clarified. Land pressures are only expected to increase across the Global South (see Chap. 7), making open and transparent legal recognition of rights all the more needed.

Understanding the nuances of how customary and formal systems align (or do not) is challenging but critical for recognizing and elevating customary lands to legitimate legal status (Knight, 2010). Even when formalization of customary and traditional lands has been attempted, the initial weak legal status of these traditional lands (especially ones that have not been put into "productive use") can leave them vulnerable to claim or co-opt by outside parties (Alden Wily, 2011). Still, it is encouraging that recognition of forests as owned or designated for Indigenous People and local communities grew steadily from 2002 to 2017 (RRI, 2018).

Moving forward, aligning *de jure* rights—the statutory and legal codes and policies that define and provide backing for rights—and *de facto* understandings of who holds which rights, and how tenure security plays out on the ground is necessary for equitable and transparent governance of land and for landholders to feel secure in making sustainable investments in land and property. In many cases, communities can leverage rights "on the books" or in the legal code. Ensuring communities know their legal rights, and have ways to monitor and enforce those rights, is also fundamental to ensuring congruence between *de jure* and *de facto* realities. In the context of our space explorer, small, isolated communities can develop *de facto* arrangements to help bring order to land and property relations within their own landscape. As populations grow and various (outside) entities develop interests in scarce land resources, governance mechanisms must develop coherent ways of allocating land to address growing populations and emerging economic activities. A state helps formalize governance, back right holdings, and adjudicate disputes. In some cases, the core role of the state may be to provide formal recognition of *de facto* right contexts.

Throughout the remainder of this book, some chapters discuss situations and attempts to formalize rights, while others grapple with understanding strategies to recognize customary or *de facto* contexts. A recognition of how "on paper" rights and "on the ground" rights can differ, as well as strategies and mechanisms for how we might align these, is one key to developing sustainable and equitable land management into the future.

References

Alden Wily, L. (2011). 'The law is to blame'. Taking a hard look at the vulnerable status of customary land rights in Africa. *Development and Change, 42*(3), 733–757. https://doi.org/10.1111/j.1467-7660.2011.01712.x

Alden Wily, L. (2018). Collective land ownership in the 21st century: Overview of global trends. *Land, 7*(2), 26. https://doi.org/10.3390/land7020068

Banner, S. (2009). *How the Indians lost their land: Law and power on the frontier*. Harvard University Press.

Bromley, D. W. (1992). *Making the commons work: Theory, practice, and policy*. ICS Press.

Bromley, D. W. (2004). Property rights: Locke, Kant, Peirce and the logic of volitional pragmatism. In H. M. Jacobs (Ed.), *Private property in the 21st century: The future of an American ideal* (pp. 19–30). Edward Elgar Publishing.

Bromley, D. W. (2008). Resource degradation in the African commons: Accounting for institutional decay. *Environment and Development Economics, 13*(05), 539–563.

Chimhowu, A. (2019). The 'new' African customary land tenure. Characteristic, features and policy implications of a new paradigm. *Land Use Policy, 81*(July 2018), 897–903. https://doi.org/10.1016/j.landusepol.2018.04.014

Cooter, R., & Ulen, T. (2012). Law and economics, 6th edition. In *6th edition*. Addison-Wesley.

Deininger, K., Ali, D. A., & Alemu, T. (2011). Impacts of land certification on tenure security, investment, and land market participation: Evidence from Ethiopia. *Land Economics, 87*(2), 312–334. http://le.uwpress.org/content/87/2/312.abstract

Demsetz, H. (1967). Toward a theory of property rights. *The American Economic Review, 57*(2), 347–359.

Diop, A. B. (1968). La Tenure Foncière En Milieu Rural Wolof (Sénégal): Historique et Actualité. *Notes Africaines: Bulletin d'information et de Correspondance de l'Institut Français d'Afrique Noire, 118*, 48–52.

Ellis, E. C., Gauthier, N., Goldewijk, K. K., Bird, R. B., Boivin, N., Díaz, S., Fuller, D. Q., et al. (2021). People have shaped Most of terrestrial nature for at least 12,000 years. *Proceedings of the National Academy of Sciences, 118*(17), e2023483118. https://doi.org/10.1073/pnas.2023483118

Executive Office of the President Council of Economic Advisers. (2021). Exploring new Frontiers in space policy and property rights. In *Economic report of the president (USA)* (pp. 225–249). U.S. Government Publishing Office.

Fenske, J. (2011). Land tenure and investment incentives: Evidence from West Africa. *Journal of Development Economics, 95*(2), 137–156. https://doi.org/10.1016/j.jdeveco.2010.05.001

Gelder, J.-L. v. (2010). What tenure security? The case for a tripartite view. *Land Use Policy, 27*(2), 449–456. https://doi.org/10.1016/j.landusepol.2009.06.008

Ghebru, H., & Lambrecht, I. (2017). Drivers of perceived land tenure (in)security: Empirical evidence from Ghana. *Land Use Policy, 66*(April), 293–303. https://doi.org/10.1016/j.landusepol.2017.04.042

Higgins, D., Balint, T., Liversage, H., & Winters, P. (2018). Investigating the impacts of increased rural land tenure security: A systematic review of the evidence. *Journal of Rural Studies, 61*(June), 34–62. https://doi.org/10.1016/j.jrurstud.2018.05.001

Ho, P. (2015). Myths of tenure security and titling: Endogenous, institutional change in China's development. *Land Use Policy, 47*(January 2013), 352–364. https://doi.org/10.1016/j.landusepol.2015.04.008

Holland, M. B., de Koning, F., Morales, M., Naughton-Treves, L., Robinson, B. E., & Suárez, L. (2014). Complex tenure and deforestation: Implications for conservation incentives in the Ecuadorian Amazon. *World Development, 55*(March), 21–36. https://doi.org/10.1016/j.worlddev.2013.01.012

Knight, Rachel. (2010). Statutory Recognition of Customary Land Rights in Africa: An Investigation into Best Practices for Lawmaking and Implementation. 105. FAO Legislative Study. Rome, Italy.

Lin, J. Y. (1992). Rural reforms and agricultural growth in China. *The American Economic Review, 82*(1), 34–51. http://www.jstor.org/stable/10.2307/2117601

Locke, J. (1689). Two treatises of government. *The Anthropology of Citizenship: A Reader, 2013*, 43–46.

Ma, X., Heerink, N., Ierland, E. v., Lang, H., & Shi, X. (2020). Decisions by Chinese households regarding renting in arable land—The impact of tenure security perceptions and trust. *China Economic Review, 60*(6), 101328. https://doi.org/10.1016/j.chieco.2019.101328

Meinzen-Dick, R., & Mwangi, E. (2009). Cutting the web of interests: Pitfalls of formalizing property rights. *Land Use Policy, 26*(1), 36–43. https://doi.org/10.1016/j.landusepol.2007.06.003

Ostrom, E. (2002). *The Drama of the commons*. National Academies Press.

Pélissier, Paul. (2008). *Les Paysans Du Senegal; Les Civilisations Agraires Du Cayor a La Casamance. Revue Tiers Monde (Vol. 8)*. Dakar-Paris.

Prindex. (2020). Security and formalization. In *Land and property rights in sub-Saharan Africa*. Prindex. https://www.prindex.net/reports/six-infographics-land-and-property-rights-sub-saharan-africa/

Robinson, B. E., Provencher, B., & Lewis, D. J. (2013). Managing wild resources: Institutional choice and the recovery of resource rent in Southwest China. *World Development, 48*(August), 120–132. https://doi.org/10.1016/j.worlddev.2013.04.004

Robinson, B. E., Masuda, Y. J., Kelly, A., Holland, M. B., Bedford, C., Childress, M., Fletschner, D., et al. (2018). Incorporating land tenure security into conservation. *Conservation Letters, 11*(2), e12383. https://doi.org/10.1111/conl.12383

RRI. (2018). At a Crossroads: Consequential Trends in Recognition of Community-Based Forest Tenure, no. September: 60.

Schlager, E., & Ostrom, E. (1992). Property-rights regimes and natural resources: A conceptual analysis. *Land Economics, 68*(3), 249–262. https://doi.org/10.2307/3146375

Sjaastad, E., & Bromley, D. W. (2000). The prejudices of property rights: On individualism, specificity, and security in property regimes. *Development Policy Review, 18*(4), 365–389.

Stickler, M. M., Huntington, H., Haflett, A., Petrova, S., & Bouvier, I. (2017). Does de facto Forest tenure affect Forest condition? Community perceptions from Zambia. *Forest Policy and Economics, 85*(1), 32–45. https://doi.org/10.1016/j.forpol.2017.08.014

Tseng, T.-w. J., Robinson, B. E., Bellemare, M. F., Yishay, A. B., Blackman, A., Boucher, T., Childress, M., et al. (2021). Influence of land tenure interventions on human well-being and environmental outcomes. *Nature Sustainability, 4*(3), 242–251. https://doi.org/10.1038/s41893-020-00648-5

Wang, H., Riedinger, J., & Jin, S. (2015). Land documents, tenure security and land rental development: Panel evidence from China. *China Economic Review, 36*, 220–235. https://doi.org/10.1016/j.chieco.2015.09.005

Wolford, W., Borras, S. M., Hall, R., Scoones, I., & White, B. (2013). Governing global land deals: The role of the state in the rush for land. *Development and Change, 44*(2), 189–210. https://doi.org/10.1111/dech.12017

4

Indigenous and Customary Land Tenure Security: History, Trends, and Challenges in the Latin American Context

Laura Aileen Sauls, Fernando Galeana, and Steven Lawry

Why Indigenous and Customary Land Tenure?

Globally, land rights advocates consistently argue that the statutory recognition of Indigenous and customary forms of tenure can encourage the sustainable management of local resources, contribute to global environmental services (such as carbon sequestration), and leverage new forms of

L. A. Sauls (✉)
Department of Geography, University of Sheffield, Sheffield, UK

Global Affairs, George Mason University, Fairfax, VA, USA
e-mail: l.a.sauls@sheffield.ac.uk

F. Galeana
Sociology and Integrative Conservation, University of William & Mary, Williamsburg, VA, USA

S. Lawry
CIFOR, Bogor, Indonesia

M. B. Holland et al. (eds.), *Land Tenure Security and Sustainable Development*,
https://doi.org/10.1007/978-3-030-81881-4_4

public and private investment that enhance livelihoods (Blackman et al., 2017; Rights and Resources Initiative, 2018a; Robinson et al., 2014). But what are these Indigenous and customary tenure forms? Where do these tenure regimes come from? And what role do they play in promoting sustainable development?

Addressing these questions requires that we first put the contemporary trend of recognizing Indigenous and customary land rights into a historical and geographical perspective. What we now call customary forms of tenure reflect the interactions over time between pre-colonial land-holding institutions with new forms of tenure introduced during the colonial period. After independence, these forms underwent transformation as part of state-building programs and continue to evolve in tandem with evolving policy priorities. Far from static and anachronistic, today's Indigenous and customary tenure regimes are dynamic systems of governance that reflect local conditions, national legal frameworks, and international policy priorities (Fitzpatrick, 2005). Because of their dynamic and often hybrid nature, these regimes resist categorization under the dominant binary of private versus public land (Ostrom, 2009). They reflect cultural and spiritual values and practices that represent distinct forms of relating to place and nature, which cannot be reduced to a statutory classification of property (Escobar, 2008).

Indigenous and customary land tenure regimes are as diverse as the landscapes and regions in which they exist, and this chapter can only provide a partial view of the core issues facing Indigenous and customary land tenure security today. That said, by highlighting the diversity and dynamism of these systems and their common differentiation from Western property systems, we aim to provide key definitions and background for understanding their status today. This chapter also analyzes contemporary trends for these regimes to highlight the vital role that such tenure regimes still play for human-environment relationships across the globe and prospects for sustainable development more broadly.

We will first define what we mean by a customary land tenure regime, explaining the particularities of making claims on the basis of indigeneity. Next, we will explain how the systems that we recognize today as Indigenous or customary have their roots in pre-colonial governance systems but co-evolved with changes during colonial times and in the post-colonial period. From there, we examine contemporary trends in Indigenous and customary land rights recognition and tenure security, with a focus on lowland tropical regions in Latin America where more extensive recognition has taken place (Rights and Resources Initiative, 2018b). Finally, we consider current and projected challenges to achieving security for communities whose livelihoods and territory depend on recognition of their customary and Indigenous land tenure regime, especially in the context of increasing pressures on their historic lands.

Defining (Customary and Indigenous) Land Tenure Regimes

According to a Rights and Resources Initiative (RRI) study of land rights in 64 countries, Indigenous Peoples and local communities hold nearly 64 percent of the world's land and have formal recognition to about 18 percent of it (RRI, 2015).[1] Other estimates suggest that these communities may have rights to up to 25 percent of the world's land area, including 40 percent of terrestrial protected areas (Garnett et al., 2018). At least 1.5 billion people depend on lands under customary or Indigenous control for their livelihoods (Rights and Resources Initiative, 2015). Many, though not all, of these areas are remote or in frontier zones, containing some of the most globally significant remaining intact forest landscapes—recent estimates suggest over 1 billion people reside within 5 kilometers of a forest (Fa et al., 2020; Newton et al., 2020).

Many communities around the world have developed collective land tenure systems to manage the lands, forests, and resources that they held in common. Over long periods of time, communities have adapted these systems to their local environments resulting in highly diverse institutional forms responsive to local ecological and cultural conditions (Cotula, 2007; Trawick, 2001; Unruh, 2006). Not all collective land tenure systems, however, are held under customary tenure. The case of the Association of Forest Communities in Petén (Guatemala) demonstrates a contemporary example where forests are collectively managed, but the state is the guarantor of and has jurisdiction over the resources (Gómez & Ernesto Mendéz,

[1] The RRI study classified different types of land tenure systems as community-based tenure regimes if ownership or management were held at the community level. Formal recognition included ownership and management rights. Several United Nations agreements also use this language to cover a range of traditional and customary land and resource rights arrangements.

2007; Sauls, 2020). This system also evolved out of negotiations between the state and communities in the context of the 1996 Central American Peace Accords, which stipulated that 100,000 ha of land be turned over to organized communities within the Maya Biosphere Reserve (Gnych et al., 2020). The Mexican *ejido* system, codified in the post-revolutionary Constitution of 1917, similarly situates a form of collective land management firmly within the state's purview, although it did grow out of customary forms of post-colonial collective land-holding (Barnes, 2009).

Customary and Indigenous land tenure regimes are those where community, rather than statutory, norms and rules usually prevail when it comes to land use decisions. In these tenure regimes, "people gain access to the commons as a social right derived from their membership in the local community or collective" (Gnych et al., 2020, 2). The ways in which people gain access are mediated through often unwritten protocols and practices that set criteria for membership and status in the territorial groups. These protocols may incorporate various conceptualizations of the rights of nature into these regimes, reflecting distinctive, place-based worldviews (Escobar, 2008; Sánchez Canseco, 2017). For this reason, the range of "different relations of authority, identity or territory" between people and land—or *land tenure regimes*—that we define as customary or Indigenous thus are important for understanding social, economic, and environmental conditions across a range of geographies (Boone, 2015, 173).

Customary land tenure regimes often incorporate a mix of ownership and use rights and are found in urban areas, too. As Chimhowu (2019, 898) notes, "customary tenure is an omnibus term that at its most basic means collectively owned land usually under the authority of traditional leadership," even if within the collectively held lands, certain plots are allotted as almost individual property. Thus, some customary regimes function even where land is no longer collectively held, but where a local community retains jurisdiction over land-community relations, as is now the case in many parts of Sub-Saharan Africa (Alden Wily, 2017). Distinct customary land tenure arrangements have important implications for urban land policy, especially in Africa where traditional authorities have jurisdiction over larger areas of urban and peri-urban land compared to other regions (Pieterse & Parnell, 2014). For an example of this phenomenon, see Box 4.1, which details the historical evolution of Mailo tenure regime in Uganda and implications for investment in Kampala, Uganda's rapidly growing capital.

Box 4.1 Mailo Customary Land Tenure System in Uganda

Present-day customary land tenure arrangements in Uganda are multi-layered, reflecting the country's history. The Kingdom of Buganda, located within the country's central region, has its own land tenure system known as Mailo. The kingdom has a history dating back to the fourteenth century (Wrigley, 1996). However, the origins of Mailo date back to the colonial period, c. 1900, when the British awarded the Buganda nobility private property rights over their feudal estates. The new system gave the nobility absolute control over the land, disregarding the historic customary use rights of the peasantry. In 1928, the legislation was amended to grant these tenants protection against rent hikes and evictions. Following Uganda's independence in 1962, the Kingdom was abolished and Mailo land was nationalized. When the National Resistance Movement came to power in 1986, royalists began to negotiate for the restoration of the Buganda monarchy. A new Kabaka (king) was crowned in 1993, and subsequently, the 1995 Constitution recognized the Mailo system (Green, 2006).

Today, Mailo tenure prevails on about 10 percent of Uganda's land area including Uganda's most urbanized and economically dynamic region, which includes the capital city of Kampala. A principal feature of the Mailo system is that absentee landlords, including descendants of the nobility, own most of the land. The Kabaka is the largest landowner and his estates are managed by the Buganda Land Board. Most land users are considered occupants. The Land Act of 1998 formalized occupancy rights for lawful tenants who could claim rights based on legacy legislation from the colonial period. In addition, a category of good faith (bona fide) tenants was created to recognize the occupancy rights of those who had held land for more than 12 years without challenge by the owners or who were part of government resettlement programs before the approval of the 1995 Constitution.

Despite legal protections, occupants on Mailo land continue to face land tenure insecurity and large-scale evictions are not uncommon (Musinguzi et al., 2020; Place & Otsuka, 2002). The Buganda Land Board has tried to improve the situation by extending certificates of occupancy to lawful and bona fide occupants. This certificate grants the occupant the right to own, sell, and mortgage improvements on the land for a given number of years (49, 75, or 99 years with the option to renew). Resolving overlapping claims is expected to give landlords and tenants more options to negotiate land agreements (Musinguzi et al., 2020). Because of the urban and economic geography of these land claims, improving the security of Mailo tenure has implications not only for landlord-tenant relations but also for the country's urban development and economic growth (Deininger & Ali, 2008). The legacy of land tenure conflicts on Mailo land led to unregulated informal development, which undermined substantial economic growth for the country (Bird & Venables, 2020).

While "customary tenure" serves as a useful term to describe land tenure rights that derive from community or collective membership, it masks the great diversity of actual institutional forms, social relations, and land use practices related to commons management (Fitzpatrick, 2005). For one thing, customary tenure regimes are local adaptations to ecosystems that vary greatly around the globe, from the highlands of South America's Andean Mountains to the dense rainforests of Kalimantan, in Indonesia, to the tundra of northern Canada. Intensive agricultural systems demand different relations between communities and their lands than do multi-local hunter-gatherer systems, so different customary tenure regimes are composed of a diverse array of norms that respond to environmental conditions as well as cultural distinction (Sikor & Stahl, 2011).

Additionally, while customary tenure regimes may function outside of statutory rules regarding land and property, the state may still play a significant role in constraining or structuring forms of customary tenure. In some contexts, the state may sanction which customary authorities are legitimate and guarantee their authority, as Boone (2015) documents in the Asante region of central Ghana. In other cases, the absence of the state—or persistent resistance to it—has helped define the institutions in an Indigenous or customary system, as is the case in the *comarcas* of Panama (Spalding, 2017). Whether the state directly sanctions a given customary land tenure regime or not, virtually all reflect the dynamic interplay with the colonial and post-colonial regimes that have sought to eliminate, assimilate, or incorporate them (Alden Wily, 2018; Monterroso & Larson, 2013). Thus, while customary tenure is an omnibus term, it can provide a useful lens through which to examine a variety of concerns related to land tenure, sustainable development, climate change, and ecological outcomes.

To conclude this section, we note that, although Indigenous land tenure regimes could be considered customary, they are granted special protection under international law—through the International Labor Organization's Indigenous and Tribal Peoples Convention 169 (ILO 169), ratified in 1989, and the United Nations Declaration on the Rights of Indigenous Peoples (UNDRIP), adopted in 2007. While UNDRIP declines to give a definition of "indigenous," ILO 169 states that it applies to self-identified groups with "descent from the populations which inhabited the country, or a geographical region to which the country belongs, at the time of conquest or colonisation or the establishment of present state boundaries and who, irrespective of their legal status, retain some or all of their own social, economic, cultural and political institutions" (International Labour Organization (ILO), 1989). These groups often, but not always, maintain distinct linguistic, religious, subsistence, and cultural practices that set them apart from the majority populations in colonized geographies, in addition to retaining their customary land tenure regimes. Further, in many cases, these groups make claims to *territorial* rights, which include secure land tenure, but also a broader array of cultural and historical recognition (López Sandoval et al., 2017).

Pre-Colonial Land Tenure and Today's Customary Systems: A View from Latin America

In what is now known as the Americas, moments of encounter between colonizing Europeans and pre-colonial societies transformed landscapes and land tenure systems—often violently. This violence still affects many landscapes where Indigenous land tenure regimes persist, as exclusion, expropriation, and extraction threaten traditional claims to land and culture (Bebbington et al., 2018; Global Witness, 2019). At the same time, the forms of tenure in Indigenous-held and -claimed regions today are in part a reaction to or function of these colonial encounters, even as they can also resist state control. These Indigenous land tenure regimes are neither relics of a pre-colonial past nor do they conform with the ideal type of individual and collectivized property that dominated land reforms in the twentieth century.

Pre-colonial Latin America contained highly varied land tenure regimes, from the extensive, bureaucratic, and agriculturally advanced Inca and Aztec empires to the loose networks of multi-local communities that populated (and in some areas, still populate) the Amazon (Angeles & Elizalde, 2017). Some of these regimes rivaled European systems of land allotment, registration, and cadaster; however, as Lockhart (1992, 162) details of the Nahua of Central Mexico, "the dominant relationship between public and private in the European tradition is that land is either one or the other, while the dominant relationship in the Nahua tradition was that it was both at the same time." Thus, while individual households might have possessed, inherited, and sold parcels of land as if they were private, the origins of the right of possession derived from membership in the collective and required approval or endorsement by collective authorities. As this author continues, post-colonization "a strong indigenous base continued to provide the framework while Spanish items and modes quickly entered everywhere, not so much displacing as infiltrating, interpenetrating, and being assigned to niches already existing in the indigenous cultural scheme" (Lockhart, 1992, 202). Today, while the Nahua customary land tenure regime has faded from view, its influence lives on in the *ejido*—which still recognizes a form of individual land use and rights nested within a collective (Olson, 2014).

Some of the earliest colonial encounters, between Spanish invaders and Caribbean societies, essentially wiped out entire Indigenous populations, after which colonizers supplanted traditional land tenure regimes with completely new forms, such as the slave-dependent plantation model. Yet even this system, a quasi-feudal import adapted to commodity agriculture for maximum profit, could not stop other forms of customary land-holding from flourishing along its periphery. The case of the *maroon* communities of Jamaica demonstrates where a customary system has emerged directly from a colonial land tenure regime, or more specifically, in fleeing from it (Zips, 1998). These communities incorporated values, language, and culture from multiple African communities, translated through enslavement and resistance to it, in a relatively new environment to produce a new collective form of small-holder-based agriculture on collectively held lands. Despite concerted efforts by British and then Jamaican authorities to stamp them out through both military and legislative campaigns, the Leeward and Windward maroons of Jamaica retain rights to the lands they claimed well over three hundred years ago and continue to maintain their customary land tenure regime and practices.

Finally, many Indigenous groups that depend upon customary land tenure regimes do so under conditions quite distinct from their pre-colonial settings, both geographically and institutionally. For example, as Le Tourneau (2015) notes, many of the Indigenous groups that now live in the Brazilian Amazon originated elsewhere and came to their current territories in the processes of fleeing colonial campaigns. The customary regimes that they now practice retain elements of their pre-colonial systems, but have also adapted to new conditions, and at times assimilated practices or peoples from other traditional groups. Further, in much of Latin America, many Indigenous communities were able to secure collective land rights during the wave of multicultural reforms that swept the region in the 1990s and 2000s (Bryan, 2012). The process of achieving these rights has been fraught, and remains incomplete; however, the recognition of customary land tenure under constitutional law, such as in Brazil, Bolivia, and Ecuador, indicates a statutory acceptance of a diversity of tenure forms—under certain conditions (Kröger & Lalander, 2016; Offen, 2003).

Trends in Customary Land Tenure Regimes and their Recognition

Globally, nearly 80 percent of all forested land designated for use by Indigenous Peoples and local communities was recognized since 2002, while Indigenous Peoples and local communities gained ownership rights over an additional 90 million ha between 2002 and 2017 (Rights and Resources Initiative, 2018b). These gains are substantial and represent both a significant shift in policy and the results of long-term struggles by Indigenous Peoples and traditional communities. That said, much more land than is recognized remains under customary use, while statutorily owned by the State or even private actors. Further, even recognized or titled customary lands remain insecure for many peoples.

This section briefly reflects on recent efforts to support the recognition of customary land rights, again drawing in particular from trends in Latin America. This experience highlights how social movements around identity and rights intersected with more market-oriented land reform efforts, resulting in hybridized models of land ownership and use rights with the

State in the role of guarantor. It also examines the link between land rights and forest and biodiversity conservation. The tensions between agrarian land reform, conservation, and recognition of customary rights also play out in other regions, but the very different contexts of colonial and post-colonial policies make for significant differences. For example, Box 4.1 provides additional insights on the formalization of occupancy rights on customary land tenure system created during the colonial period in Uganda.

The move from the late 1980s onward toward the recognition of collective land rights and customary land tenure regimes reflected a sharp break with previous agrarian reform policies, especially in Latin America (Pacheco et al., 2012). In previous decades, land reform generally meant the re-distribution of lands from large landholders (in Latin America, a remnant of the *hacienda* system) and allocation of parcels or plots to individual families through the designated head of household. Land reform that supported rights for small-holder farmers—*campesinos* in Latin America—were fundamental markers of progressive regimes in the twentieth century, though they met with variable success. Later efforts to improve land registration, especially market-led plans that would support burgeoning land markets and investment, attracted significant investment from international donors, including the World Bank (Hetherington, 2012). Accompanying these processes were often technocratically driven large-scale cadastral surveys and registration projects to make legal and formalize these new property relations (Fontana, 2014). These processes are covered in greater detail in Chaps. 3 and 11 of this volume.

The reactivation of indigeneity as a political identity starting in the 1970s pushed agrarian reform to consider multicultural policies that recognized demands for territory more broadly (Van Cott, 2000). These demands by Indigenous Peoples as well as other traditionally marginalized communities, such as Afro-descendant and forest-dependent groups, in part responded to the rapid expansion of the agricultural frontier, which threatened their access to land as well as the customary tenure regimes that guided their relation to it. In Latin America, rising concern on the part of environmentalists over the loss of biodiversity and tropical forests along these new frontiers coincided with community-based land rights demands. The combination of these concerns drew renewed attention to the loss of fragile ecosystems and distinct socio-environmental relations, and bolstered international attention to the loss of forests, including those under Indigenous and customary tenure (Keck & Sikkink, 1998). While for many groups, ethnic identity played a central role in their land tenure demands, the pairing of conservation and community concern over the loss of tropical forests often played an important role in encouraging the recognition of customary or collective systems for non-Indigenous identifying groups, such as the rubber tapping communities of Acre, Brazil (Rodrigues, 2015). At the same time, Indigenous control may not automatically result in improved forest outcomes, and some Indigenous communities have had tense relations with conservation-focused organizations that advocate for limits on community exercise of land use rights (Holland et al., 2017; Lu et al., 2010).

Thus, the intersection of demands by social movements for collective land rights, international concern for the environment, and policy responses to improve land governance in support of sustainable development fundamentally shifted land rights processes across Latin America, as well as other parts of the Global South. While these processes have remade statutory land tenure systems, their effects on Indigenous and customary land tenure remains less clear. Box 4.2 provides a case where these different demands intersected in efforts to formalize customary land tenure among the Miskitu people in Honduras and Nicaragua.

Box 4.2 Titling and Miskitu Land Tenure in Honduras and Nicaragua

The Muskitia, which encompasses a significant portion of the Caribbean coasts of Northeastern Honduras and Nicaragua, is the binational homeland of the Miskitu people. During the colonial period, the Miskitu managed an autonomous kingdom in alliance with the British against the incursion of Spanish Central America. Although the Miskitu Kingdom dissolved in 1860, the region has maintained its territorial congruity and a distinctive cultural-linguistic identity (Pineda, 2006).

Like many other Indigenous Peoples in tropical forests, the Miskitu practice agriculture, hunting, and fishing as part of their subsistence economy. The customary land tenure system is based on the traditional occupation of extensive areas by kinship groups (known as kiamka), which hold land collectively over generations. In these lands, the Miskitu practice swidden agriculture, rotating about every two years, allowing the nutrient-poor tropical

(continued)

Box 4.2 (continued)

soils to regenerate. Many households must travel long distances to reach their farming plots, some of which are left fallow for decades but still recognized as part of a *kiamka*'s trust.

Until recently, state policies in Honduras and Nicaragua dismissed that this complex land tenure system could be the basis for property rights and viewed the territories occupied by the Miskitu as public lands, a situation that encouraged the migration of landless *mestizo* farmers into these areas (Finley-Brook, 2016; Galeana, 2020). As these land invasions intensified in the 1970s and 1980s, the Miskitu, like many other Indigenous Peoples across the Americas, began demanding the legal protection of their ancestral territories. In both Honduras and Nicaragua, participatory mapping projects proved to be valuable instruments in raising awareness among state officials about the territorial dimension of the Miskitu land tenure system (Herlihy & Knapp, 2003). Based on participatory cartography, the Miskitu have obtained property titles to 18 territories in Nicaragua and 12 territorial councils in Honduras during the 2000s and 2010s. In both countries, the World Bank supported the titling process as part of broader efforts to rationalize land tenure to facilitate governance and economic growth (Hale, 2011). There are jurisdictional differences, though, as the Nicaraguan state recognizes political autonomy over two regions in the Caribbean Coast, whereas the Honduran state does not.

Despite the property titles, communities on both sides of the border continue to face land tenure insecurity. The Honduran and Nicaraguan governments have both failed to provide communities with the financial and logistical support to evict non-Indigenous settlers. As a result, the titles have not stopped the intensification of land grabbing for cattle ranching, timber, and mining, which has also increased deforestation rates (McSweeney et al., 2014). Land grabbing has been notoriously violent in Nicaragua, partly as a result of the cooptation of Indigenous authorities by the political establishment (Mittal, 2020). These governance issues underscore the point that land tenure security does not automatically result from obtaining a land title but depends on the broader political and economic conditions that shape land control (Ribot & Peluso, 2003).

Despite the newfound support to the formalization of customary land tenure, in practice, the dominant economic model continues to undermine the prospects for genuine recognition of territorial rights. In many countries, efforts to title the collective land rights of Indigenous and Afro-descendant communities have taken place alongside plans for large-scale development and commodification of natural resources

(Hale, 2011; Offen, 2003). Although organizations like the World Bank have readjusted their policies to accommodate customary land rights, they continue to plan and implement programs based on economic and market principles that often exclude political rights (Anthias & Radcliffe, 2015). Therefore, the seeming policy consensus on the need to formalize customary land tenure in Latin American is built on a paradoxical arrangement of grassroots demands, conservation efforts, and economic policies. The exact details of this paradox are beyond the scope of this chapter and have been well covered by a range of scholars (cf. Anthias & Radcliffe, 2015; Hale, 2011; Mollett, 2013; Offen, 2003). Nonetheless, we recognize that this policy shift would not have taken place without the activism of Indigenous movements, which in spite of challenges across scales continue to adapt their tools and strategies in mobilizing a territorial agenda (Sauls, 2020).

Challenges to Securing Land Tenure for Customary and Indigenous Regimes

While titling customary land rights remains a priority around the world, even land titles cannot guarantee tenure security for local communities. The challenges of customary tenure post-title are as diverse as the regimes that term describes; however, three challenges are perhaps most prevalent, especially from the Latin American perspective. These include weak enforcement of protections for customary tenure rights even post-recognition or titling; changing contexts around the institutions and practices that make up customary regimes; and the rollback of protections for these regimes in the face of economic development opportunities, especially related to natural resources extraction and infrastructure expansion.

Early proponents of Indigenous and customary rights focused primarily on the need for formal recognition of historical land tenure regimes, such as via titling processes; however, subsequent events quickly demonstrated that a title has little meaning without measures to support the security of land tenure, especially on the part of a national government (Pacheco et al., 2012). Evidence from across Latin America suggests that many governments lack either the capacity or the incentive to enforce the rights bestowed by a title (Correia, 2019; Dest, 2020; Ferrante et al., 2020; McSweeney, 2020). While these rights vary depending on the country and the form of the recognition mechanism, in general, titles for territories or customary lands would exclude non-community members from extracting natural resources without permission, at the very least—and in some cases would enable community members to exclude others from accessing or using the land (Finley-Brook, 2007; Larson, 2011). As the ongoing violence against Miskitu communities in Nicaragua (Box 4.2) suggests, if the government does not make clear to broader constituencies that different rules prevail in a given territory—and that the government will enforce those rules when challenged—then community members have little recourse when outsiders—or insiders—break those rules. A tendency toward illegal land sales, land invasions, and in many cases violence in post-titling areas speak to this challenge.

On the second point, a given customary tenure regime is made up of institutions and practices grounded in a particular place, but it is also nested within broader contexts, such as the state or the globalized economy, which have their own institutions and operating logics (Alcorn & Toledo, 1995). When the broader context changes, it can undermine the functioning of the institutions that compose a given tenure regime. Those changes may result from political, economic, or environmental shifts that can alter relations within a tenure regime as well as the relations between customary norms and the institutional contexts in which they are nested. The experience of the Miskitu people (Box 4.2) outlines how some of these changes have affected the Miskitu people in Honduras and Nicaragua; other authors have also found that the requirements for receiving and maintaining tenure can shift relations of power in unexpected ways in a range of contexts (Finley-Brook, 2016; Humphreys Bebbington & Bebbington, 2010; Pacheco et al., 2012). Environmental challenges like climate change and biodiversity loss can also affect customary tenure regimes. These regimes have evolved in specific places predicated on specific socio-environmental relations, but as ecological conditions change, institutions may no longer function to effectively allocate resources—especially where movement is no longer a possibility given the rigidity of legal land rights. Lastly, the efforts to map and title customary land tenure can themselves result in the over-simplification of dynamic systems, with potential negative consequences for the future regulation of resource access and management (Bryan, 2009; Mollett, 2013).

Finally, the legibility and legalization of new territories under Indigenous or customary tenure can draw the attention of actors interested in natural resource or other economic development opportunities, which can then provide incentives to the very governments that granted those land rights to roll back protections or legislate loopholes. Indigenous territories count as a category of "protected areas" under the World Database on Protected Areas system, and recent research suggests a rising vulnerability for these systems (Forrest et al., 2015; Pack et al., 2016). Mining, oil and gas, logging, and infrastructure expansion—including for projects pitched as "low-emissions" or "green"—all require extensive land and natural resources in order to advance. In many cases, governments retain rights to sub-soil resources even where they recognized Indigenous or community rights, meaning that communities may have little recourse when a company arrives on the scene to begin drilling or mining (Bebbington et al., 2018). Further, projects deemed "in the national interest" may trump existing land rights; governments have consistently deemed large-scale infrastructure projects as worth this type of designation and ignored or shifted regulations to promote them (Kröger & Lalander, 2016; Le Tourneau, 2015). Especially in Latin America, lands under Indigenous and customary tenure regimes are also consistently among the most biodiverse and most threatened; undermining Indigenous and customary systems also undermines these fragile ecosystems and the socio-ecological systems that have helped define them (Blackman et al., 2017; Fa et al., 2020).

Conclusion

At the beginning of this chapter, we set out to define what Indigenous and customary tenure regimes are, how these systems evolved over time, and the role they play in encouraging sustainable development. We described Indigenous and customary land tenure regimes as those where community norms rather than statutory rules prevail when it comes to land use decisions. With a focus on Latin America, we placed today's customary systems within historical context, particularly highlighting the importance of colonial encounters, formalization programs, and international influences (both formal and via social movements) in shaping contemporary land tenure in historically Indigenous and customary areas. The trend toward recognition of Indigenous and customary land rights marked a shift in national land governance policy as well as a growing global recognition of the importance of both tropical forests and ecosystems and Indigenous rights; however, tenure insecurity remains a major challenge for communities that hold land through Indigenous or customary tenure systems. This insecurity results from a combination of factors—weak enforcement, changing conditions for land tenure, and new interests in land and natural resources—all of which potentially undermine the potential for sustainable development for people living under Indigenous or customary tenure regimes.

References

Alcorn, Janice B., & Toledo, Víctor M. (1995). The Role of Tenurial Shells in Ecological Sustainability Property Rights and Natural Resource Management in Mexico. In *Reinventing the Commons.*Brode, Norway. http://dlc.dlib.indiana.edu/dlc/bitstream/handle/10535/2062/The_Role_of_Tenurial_Shells_in_Ecological_Sustainability_Property_Rights_and_Natural_Resource_.pdf?sequence=1 & isAllowed=y.

Alden Wily, L. (2017). Customary tenure: Remaking property for the 21st century. In *Comparative property law: Global perspectives*. Edward Elgar Publishing. https://www.elgaronline.com/view/edcoll/9781848447578/9781848447578.00030.xml

Alden Wily, L. (2018). Collective land ownership in the 21st century: Overview of global trends. *Land, 7*(2), 68. https://doi.org/10.3390/land7020068

Angeles, L., & Elizalde, A. (2017). Pre-colonial institutions and socioeconomic development: The case of Latin America. *Journal of Development Economics, 124*(January), 22–40. https://doi.org/10.1016/j.jdeveco.2016.08.006

Anthias, P., & Radcliffe, S. A. (2015). The ethno-environmental fix and its limits: Indigenous land titling and the production of not-quite-neoliberal natures in Bolivia. *Geoforum, 64*(August), 257–269. https://doi.org/10.1016/j.geoforum.2013.06.007

Barnes, G. (2009). The evolution and resilience of community-based land tenure in rural Mexico. *Land Use Policy, 26*(2), 393–400. https://doi.org/10.1016/j.landusepol.2008.05.007

Bebbington, A. J., Bebbington, D. H., Sauls, L. A., Rogan, J., Agrawal, S., Gamboa, C., Imhof, A., et al. (2018). Resource extraction and infrastructure threaten Forest cover and community rights. *Proceedings of the National Academy of Sciences, 115*(52), 13164–13173. https://doi.org/10.1073/pnas.1812505115

Bird, J., & Venables, A. J. (2020). Land tenure and land-use in a developing City: A quantitative spatial model applied to Kampala, Uganda. *Journal of Urban Economics, 119*(September), 103268. https://doi.org/10.1016/j.jue.2020.103268

Blackman, A., Corral, L., Lima, E. S., & Asner, G. P. (2017). Titling indigenous communities protects forests in the Peruvian Amazon. *Proceedings of the National Academy of Sciences, 114*(16), 4123–4128. https://doi.org/10.1073/pnas.1603290114

Boone, C. (2015). Land tenure regimes and state structure in rural Africa: Implications for forms of resistance to large-scale land acquisitions by outsiders. *Journal of Contemporary African Studies, 33*(2), 171–190. https://doi.org/10.1080/02589001.2015.1065576

Bryan, J. (2009). Where would we be without them? Knowledge, space and power in indigenous politics. *Futures, Futures of Indigenous Knowledges, 41*(1), 24–32. https://doi.org/10.1016/j.futures.2008.07.005

Bryan, J. (2012). Rethinking territory: Social justice and neoliberalism in Latin America's territorial turn. *Geography Compass, 6*(4), 215–226. https://doi.org/10.1111/j.1749-8198.2012.00480.x

Chimhowu, A. (2019). The 'new' African customary land tenure. Characteristic, features and policy implications of a new paradigm. *Land Use Policy, 81*, 897–903. https://doi.org/10.1016/j.landusepol.2018.04.014

Correia, J. E. (2019). Unsettling territory: Indigenous mobilizations, the territorial turn, and the limits of land rights in the Paraguay-Brazil borderlands. *Journal of Latin American Geography, 18*(1), 11–37. https://muse.jhu.edu/article/719690

Cotula, L. (Ed.). (2007). *Changes in customary land tenure Systems in Africa*. International Institute for Environment and Development and Food and Agriculture Organization.

Deininger, K., & Ali, D. A. (2008). Do overlapping land rights reduce agricultural investment? Evidence from Uganda. *American Journal of Agricultural Economics, 90*(4), 869–882. https://doi.org/10.1111/j.1467-8276.2008.01171.x

Dest, A. (2020). 'Disenchanted with the state': Confronting the limits of neoliberal multiculturalism in Colombia. *Latin American and Caribbean Ethnic Studies, 1*, 1–23. https://doi.org/10.1080/17442222.2020.1777728

Escobar, A. (2008). *Territories of difference: Place, movements, life, Redes*. Duke University Press.

Fa, J. E., Watson, J. E. M., Leiper, I., Potapov, P., Evans, T. D., Burgess, N. D., Molnár, Z., et al. (2020). Importance of indigenous peoples' lands for the conservation of intact Forest landscapes. *Frontiers in Ecology and the Environment, 18*(3). https://doi.org/10.1002/fee.2148

Ferrante, L., Gomes, M., & Fearnside, P. M. (2020). Amazonian indigenous peoples are threatened by Brazil's highway BR-319. *Land Use Policy, 94*(May), 104548. https://doi.org/10.1016/j.landusepol.2020.104548

Finley-Brook, M. (2007). Indigenous land tenure insecurity fosters illegal logging in Nicaragua. *International Forestry Review, 9*(4), 850–864. https://doi.org/10.1505/ifor.9.4.850

Finley-Brook, M. (2016). Territorial 'fix'?Tenure insecurity in titled indigenous territories. *Bulletin of Latin American Research, 35*(3), 338–354. http://onlinelibrary.wiley.com/doi/10.1111/blar.12489/full

Fitzpatrick, D. (2005). 'Best practice' options for the legal recognition of customary tenure. *Development and Change, 36*(3), 449–475. https://doi.org/10.1111/j.0012-155X.2005.00419.x

Fontana, L. B. (2014). Indigenous peoples vs peasant unions: Land conflicts and rural movements in Plurinational Bolivia. *The Journal of Peasant Studies, 41*(3), 297–319. https://doi.org/10.1080/03066150.2014.906404

Forrest, J. L., Mascia, M. B., Pailler, S., Abidin, S. Z., Araujo, M. D., Krithivasan, R., & Riveros, J. C. (2015). Tropical deforestation and carbon emissions from protected area downgrading, downsizing, and Degazettement (PADDD). *Conservation Letters, 8*(3), 153–161. https://doi.org/10.1111/conl.12144

Galeana, F. (2020). Legitimating the state and the social movement: Clientelism, brokerage, and collective land rights in Honduras. *Journal of Latin American Geography, 19*(4), 11–42. https://doi.org/10.1353/lag.2020.0099

Garnett, S. T., Burgess, N. D., Fa, J. E., Fernández-Llamazares, Á., Molnár, Z., Robinson, C. J., Watson, J. E. M., et al. (2018). A spatial overview of the global importance of indigenous lands for conservation. *Nature Sustainability, 1*(7), 369–374. https://doi.org/10.1038/s41893-018-0100-6

Global Witness. (2019). *Enemies of the state?How governments and business silence land and environmental defenders.* Global Witness. https://www.globalwitness.org/sv/campaigns/environmental-activists/enemies-state/

Gnych, S., Lawry, S., McLain, R., Monterroso, I., & Adhikary, A. (2020). Is community tenure facilitating Investment in the Commons for inclusive and sustainable development? *Forest Policy and Economics, 111*(February), 102088. https://doi.org/10.1016/j.forpol.2019.102088

Gómez, I., & Ernesto Mendéz, V. (2007). *El Caso de La Asociación de Comunidades Forestales de Petén (ACOFOP): Análisis de Contexto.* PRISMA. http://www.cifor.org/publications/pdf_files/books/bcifor0802.pdf

Green, E. D. (2006). Ethnicity and the politics of land tenure reform in Central Uganda. *Commonwealth & Comparative Politics, 44*(3), 370–388. https://doi.org/10.1080/14662040600997148

Hale, C. R. (2011). Resistencia Para que?Territory, autonomy and neoliberal entanglements in the 'empty spaces' of Central America. *Economy and Society, 40*(2), 184–210. https://doi.org/10.1080/03085147.2011.548947

Herlihy, P. H., & Knapp, G. (2003). Maps of, by, and for the peoples of Latin America. *Human Organization, 62*(4), 303–314. https://www.jstor.org/stable/44127812

Hetherington, K. (2012). Promising information: Democracy, development, and the remapping of Latin America. *Economy and Society, 41*(2), 127–150. https://doi.org/10.1080/03085147.2011.607365

Holland, M. B., Dupre, S. I., Baker, J., Lawry, S., Duchelle, A. E., Kelly, A., Masuda, Y. J., & Robinson, B. E. (2017). How Does Indigeneity Influence Socio-Ecological Conditions in Community Forests?: A Systematic Review Protocol. Working Paper 234.Bogor, Indonesia: CIFOR. https://www.cifor.org/publications/pdf_files/WPapers/WP234Holland.pdf.

Humphreys Bebbington, D., & Bebbington, A. J. (2010). Extraction, territory, and inequalities: Gas in the Bolivian Chaco. *Canadian Journal of Development Studies/Revue Canadienne d'études Du Développement, 30*(1–2), 259–280. http://www.tandfonline.com/doi/abs/10.1080/02255189.2010.9669291

International Labour Organization (ILO). (1989). *Indigenous and Tribal Peoples Convention.* http://www.ilo.org/dyn/normlex/en/f?p=NORMLEXPUB:12100:0::NO::P12100_INSTRUMENT_ID:312314.

Keck, M. E., & Sikkink, K. (1998). *Activists beyond Borders: Advocacy networks in international politics*. Cornell University Press.
Kröger, M., & Lalander, R. (2016). Ethno-territorial rights and the resource extraction boom in Latin America: Do constitutions matter? *Third World Quarterly, 37*(4), 682–702. https://doi.org/10.1080/01436597.2015.1127154
Larson, A. M. (2011). Forest tenure reform in the age of climate change: Lessons for REDD+. *Global Environmental Change, 21*(2), 540–549. https://doi.org/10.1016/j.gloenvcha.2010.11.008
Le Tourneau, F.-M. (2015). The sustainability challenges of indigenous territories in Brazil's Amazonia. *Current Opinion in Environmental Sustainability, 14*, 213–220. https://doi.org/10.1016/j.cosust.2015.07.017
Lockhart, J. (1992). *The Nahuas after the conquest a social and cultural history of the Indians of Central Mexico, sixteenth through eighteenth centuries James Lockhart. ACLS humanities E-book*. Stanford University Press.
López Sandoval, M. F., Robertsdotter, A., & Paredes, M. (2017). Space, power, and locality: The contemporary use of Territorio in Latin American geography. *Journal of Latin American Geography, 16*(1), 43–67. https://doi.org/10.1353/lag.2017.0009
Lu, F., Gray, C., Bilsborrow, R. E., Mena, C. F., Erlien, C. M., Bremner, J., Barbieri, A., & Walsh, S. J. (2010). Contrasting colonist and indigenous impacts on Amazonian forests. *Conservation Biology, 24*(3), 881–885. https://doi.org/10.1111/j.1523-1739.2010.01463.x
McSweeney, K. (2020). Cocaine trafficking and the transformation of central American Frontiers. *Journal of Latin American Geography, 19*(3), 159–166. https://doi.org/10.1353/lag.2020.0075
McSweeney, K., Nielsen, E. A., Taylor, M. J., Wrathall, D. J., Pearson, Z., Wang, O., & Plumb, S. T. (2014). Drug policy as conservation policy: Narco-deforestation. *Science, 343*(6170), 489–490. https://doi.org/10.1126/science.1244082
Mittal, A. (2020). *Nicaragua's failed revolution: The indigenous struggle for Saneamiento*. The Oakland Institute. https://www.oaklandinstitute.org/sites/oaklandinstitute.org/files/nicaraguas-failed-revolution.pdf
Mollett, S. (2013). Mapping deception: The politics of mapping Miskito and Garifuna space in Honduras. *Annals of the Association of American Geographers, 103*(5), 1227–1241. https://doi.org/10.1080/00045608.2013.770366
Monterroso, I., & Larson, A. M. (2013). The dynamic Forest commons of Central America: New directions for research. *Journal of Latin American Geography, 12*(1), 87–110. https://doi.org/10.1353/lag.2013.0006

Musinguzi, M., Huber, T., Kirumira, D., & Drate, P. (2020). Assessment of the land inventory approach for securing tenure of lawful and Bona fide occupants on private Mailo land in Uganda. *Land Use Policy, 1*(March), 104562. https://doi.org/10.1016/j.landusepol.2020.104562

Newton, P., Kinzer, A. T., Miller, D. C., Oldekop, J. A., & Agrawal, A. (2020). The number and spatial distribution of Forest-proximate people globally. *One Earth, 3*(3), 363–370. https://doi.org/10.1016/j.oneear.2020.08.016

Offen, K. H. (2003). The territorial turn: Making black territories in Pacific Colombia. *Journal of Latin American Geography, 2*(1), 43–73. http://www.jstor.org.proxy-remote.galib.uga.edu/stable/25765046

Olson, E. A. (2014). *Indigenous knowledge and development: Livelihoods, health experiences, and medicinal plant knowledge in a Mexican biosphere reserve.* Lexington Books.

Ostrom, E. (2009). *Understanding institutional diversity.* Princeton University Press.

Pacheco, P., Barry, D., Cronkleton, P., & Larson, A. (2012). The recognition of Forest rights in Latin America: Progress and shortcomings of Forest tenure reforms. *Society & Natural Resources, 25*(6), 556–571. https://doi.org/10.1080/08941920.2011.574314

Pack, S. M., Ferreira, M. N., Krithivasan, R., Murrow, J., Bernard, E., & Mascia, M. B. (2016). Protected area downgrading, downsizing, and Degazettement (PADDD) in the Amazon. *Biological Conservation, 197*(May), 32–39. https://doi.org/10.1016/j.biocon.2016.02.004

Pieterse, E., & Parnell, S. (2014). Africa's urban revolution in context. In S. Parnell & E. Pieterse (Eds.), *Africa's urban revolution* (pp. 1–18). Zed Books Ltd..

Pineda, B. L. (2006). *Shipwrecked identities: Navigating race on Nicaragua's Mosquito Coast.* Rutgers University Press.

Place, F., & Otsuka, K. (2002). Land tenure systems and their impacts on agricultural investments and productivity in Uganda. *The Journal of Development Studies, 38*(6), 105–128. https://doi.org/10.1080/00220380412331322601

Ribot, J. C., & Peluso, N. L. (2003). A theory of access. *Rural Sociology, 68*(2), 153–181. https://doi.org/10.1111/j.1549-0831.2003.tb00133.x

Rights and Resources Initiative. (2015). *Who owns the World's land? A global baseline of formally recognized indigenous and community land rights.* RRI.

Rights and Resources Initiative. (2018a). *A global baseline of carbon storage in collective lands: Indigenous and local community contributions to climate change mitigation.* Rights and Resources Initiative. https://rightsandresources.org/wp-content/uploads/2018/09/A-Global-Baseline_RRI_Sept-2018.pdf

Rights and Resources Initiative. (2018b). *At a crossroads: Consequential trends in recognition of community-based Forest tenure from 2002–2017*. Rights and Resources Initiative. https://rightsandresources.org/wp-content/uploads/2018/09/At-A-Crossroads_RRI_Sept-2018.pdf

Robinson, B. E., Holland, M. B., & Naughton-Treves, L. (2014). Does secure land tenure save forests? A meta-analysis of the relationship between land tenure and tropical deforestation. *Global Environmental Change, 29*(November), 281–293. https://doi.org/10.1016/j.gloenvcha.2013.05.012

Rodrigues, M. G. (2015). Bringing local voices to the global negotiation table: Norm dissemination and consensus building on tropical forests and climate change. *New Global Studies, 9*(2), 1. https://doi.org/10.1515/ngs-2015-0016

Sánchez Canseco, J. L. (2017). La cosmovisión indígena del territorio: defensa e identidad. *Fuimos Peces, 1*(3), 1. https://www.fuimospeces.mx/single-post/2017/09/29/indigena-territorio

Sauls, L. A. (2020). Becoming fundable? Converting climate justice claims into climate finance in Mesoamerica's forests. *Climatic Change, 161*, 307–325. https://doi.org/10.1007/s10584-019-02624-1

Sikor, T., & Stahl, J. (2011). *Forests and people: Property, governance, and human rights*. Routledge.

Spalding, A. K. (2017). Exploring the evolution of land tenure and land use change in Panama: Linking land policy with development outcomes. *Land Use Policy, 61*(February), 543–552. https://doi.org/10.1016/j.landusepol.2016.11.023

Trawick, P. B. (2001). Successfully governing the commons: Principles of social Organization in an Andean Irrigation System. *Human Ecology, 29*(1), 1–25. https://doi.org/10.1023/A:1007199304395

Unruh, J. D. (2006). Land tenure and the 'evidence landscape' in developing countries. *Annals of the Association of American Geographers, 96*(4), 754–772. https://doi.org/10.1111/j.1467-8306.2006.00515.x

Van Cott, D. L. (2000). *The friendly liquidation of the past: The politics of diversity in Latin America*. University of Pittsburgh Press.

Wrigley, C. (1996). *Kingship and state: The Buganda dynasty*. Cambridge University Press.

Zips, W. (1998). 'We are landowners': Territorial autonomy and land tenure in the Jamaican maroon Community of Accompong. *The Journal of Legal Pluralism and Unofficial Law, 30*(40), 89–121. https://doi.org/10.1080/07329113.1998.10756499

5

Championing Women's Tenure Security

Diana Fletschner, Shipra Deo, and Monica Mhoja

Why Should We Care About Women's Tenure Security?

As other chapters have argued, tenure security can mean the difference between a safe and stable home and being homeless; between starting and growing a business or not; between growing long-term crops, planting trees, adopting conservation measures, and protecting the environment, or thinking only of short-term outputs; between obtaining working capital and investing in better livelihoods or being trapped in high-risk-low-return options; and between having a voice and being heard or being marginalized, ignored, or humiliated (Meinzen-Dick et al., 2019; World Bank, 2008; FAO, 2002).

D. Fletschner (✉)
Landesa, Seattle, WA, USA
e-mail: dianaf@landesa.org

S. Deo
Landesa, Lucknow, India

M. Mhoja
Landesa, Dar es Salaam, Tanzania

M. B. Holland et al. (eds.), *Land Tenure Security and Sustainable Development*,
https://doi.org/10.1007/978-3-030-81881-4_5

Yet, too often, when we discuss gaps in tenure security or put in place interventions to strengthen it, we adopt a gender-blind approach that is anchored on households, or on "heads of households," who are typically male, ignoring in both cases the complex web of gender norms and family dynamics that severely and systematically weaken women's tenure security.

We continue to use a gender-blind approach even though in large portions of the world women are consistently denied access and rights to land; even though the additional tenure insecurity increases women's physical, social and economic vulnerability; even though there is ample evidence showing that gender-blind interventions will not sufficiently address women's tenure constraints; and even though we have consistently found that interventions that enhance women's access to and control over resources are likely to lead to better outcomes for women, their families and their communities.

In Large Portions of the World Women Are Systematically Denied Access and Rights to Land

In societies where access to land is largely driven by family relations, women's ability to obtain and retain access to a plot is directly tied to the wishes of their husbands, fathers, brothers, sons, or in-laws (Lastarria-Cornhiel et al., 2014; OHCHR, 2012; Benschop, 2004). Daughters are prevented from inheriting parental land if they are generally viewed as "transitory" members of their natal household who will likely move to their husbands' household upon marriage. Wives are disposed from the land if their husbands abandon them, divorce them, or marry other wives. And women lose their land as they get older and outlive their husbands, a not-so-uncommon occurrence given women's longer life expectancy, because they become victims of land grabs through violence and bullying from their in-laws or others in their communities.

But women's *de facto* discrimination is not limited to family dynamics (Archambault & Zoomers, 2015; Lastarria-Cornhiel et al., 2014; OHCHR, 2012). Formal and customary laws often prevent women from owning, inheriting, or transacting land. Women continue to be left out of government-sponsored land allocations, land reforms, or land rights

Widowhood, Witchcraft, and Tenure Insecurity

In Northern Tanzania, one of the most harmful accusations waged against a widow is responsibility for the death of her husband. Flora's husband died of HIV, and during his burial ceremony in a village some 300 miles west of their home in Dar-es-Salaam, Flora's in-laws accused her of bewitching her husband and causing his death.

She cried all night, requesting to return to Dar es Salaam with her children. Her father-in-law wouldn't allow it, forcing her to leave without them.

When she arrived home, Flora found that her father-in-law had locked her matrimonial house, denying her access to it. Her ownership of land and property had shaped Flora's social identity, and without these assets she was insecure and powerless.

Flora's story describes what a number of widows in Tanzania and other African countries experience. Although accusations of witchcraft are not the only reason widows are left landless or impoverished, they can be a driving factor in justifying land grabbing in the eyes of those responsible, and indeed, by the community as a whole.

formalization programs that assign and document rights to the head of the household, typically a man. Women are often limited in their access to institutions or authorities that are in charge of resolving all land disputes, but whose location, staffing, processes, or cost unintentionally favors men.

Even when markets could offer women a promising path to overcome these obstacles and achieve tenure security, they may fall short of what is needed. Social norms that restrict the economic activities deemed suitable for women, financial institutions that limit women's access to working capital, and land markets biased against women all combine to constrain women's economic ability to access and retain land (Lastarria-Cornhiel et al., 2014).

Tenure Insecurity Increases Women's Physical, Social, and Economic Vulnerability

Mindful of their tenuous tenure security, women may forego livelihood strategies that could be more profitable, safe, and sustainable, because they cannot count on the long-term tenure security required to pursue those strategies; or they may grow fearful and compelled to endure physical, emotional, or economic violence because of their weak fall-back position. Since these tenure vulnerabilities are internalized and externally reinforced by society at large, they have a negative impact not just on the women who are threatened by their families, but on *all* women, because women are aware that they could one day find themselves in similar situations. The consequences are enormous for the women themselves, their families, and their communities.

Women Landowners and Leasing

Poonam, from Uttar Pradesh, India, had a happy and prosperous family. Her husband was managing their land, working in the city, and helping her raise their children. But two years ago, he was killed in a road accident.

Agriculture is now the only source of income for Poonam and her children. However, as a Brahmin, Poonam cannot cultivate the land herself, and she cannot talk to men she is not related to because of gendered social restrictions. So she is dependent on her husband's brother to lease out her land.

Her tenant is honest and hardworking, yet Poonam believes more entrepreneurial tenants would generate higher yields and thus offer her a higher share. However, Poonam feels she cannot challenge her brother-in-law's choice without compromising his willingness to assist her in the many other circumstances in which she will need the intermediation of a man.

Poonam's concerns are compounded by her lack of experience. Her husband used to be the one in charge of leasing their agricultural land and all the decisions related to it. Poonam was not privy to the considerations that went into those decisions and worries that she may not be able to find a dependable tenant by herself.

Poonam's story underscores how women, even when they do own land and belong to the highest caste, can face challenges exercising their land rights and, ultimately, become tenure insecure.

Gender-Blind Interventions Will Not Sufficiently Address Women's Tenure Constraints

Unless they specifically seek to address the gender gap, interventions to secure tenure may fail to address women's most pressing tenure constraints; may deliver goods, services or information in ways that are not helpful to women; or may fail to reach entire categories of women.

Examples of these unintended consequences abound. They include countries' efforts to document land rights using forms that only allow for one name and thereby assign rights to household heads or household representatives; systemic formalization efforts that only register rights of legally married spouses, leaving behind women in customary unions, in polygamous marriages, or who have been cohabitating with their partners for years; trainings that are held in places women cannot access or at times women are not available; information campaigns that fail to consider that women may not be fluent in the official languages or may not have direct access to common channels of communication; compensation and relocation programs that ignore women's livelihoods and their economic contributions to their families; and efforts that ignore how cultural norms and family dynamics hamper women's tenure security.

Interventions That Enhance Women's Access to and Control Over Resources Are Likely to Lead to Better Outcomes for Women, Their Families, and Their Communities

A recent study by Meinzen-Dick et al. (2019) finds that while the breadth and rigor of the evidence varies, there is a high level of agreement among development practitioners linking enhanced women's land rights to improved women's resilience, women's empowerment, women's decision-making, families' and children's food security, families' investments in human capital, and families' investments in natural resource management.

Examples often cited in the literature connect strengthened women's land rights to households' adoption of soil conservation practices in Uganda and Zambia (Deininger et al., 2008; Dillon & Voena, 2018); women making household decisions in Tanzania, India, and Nepal (Grabe, 2015; Santos et al., 2014; Allendorf, 2007); women renting out land and increasing their income in Ethiopia (Holden et al., 2011); households having more food available in Ethiopia and fewer malnourished children in Nepal (Ghebru & Holden, 2013; Allendorf, 2007); and women speaking up in community meetings, having more access to customary authorities, and having stronger social relations in Tanzania (Goldman et al., 2016; Grabe, 2015).

In fact, the foundational and cross-cutting role of women's tenure security was officially acknowledged and widely recognized in 2015 when governments across the world tied women's tenure security to 3 of the 17 Sustainable Development Goals (SDGs) they pledged to achieve by 2030: end poverty (Goal 1), achieve food security (Goal 2), and empower women and eliminate gender inequality (Goal 5). Their specific commitments with regard to women's land rights are stipulated under SDG Targets 1.4, 2.3, and 5.a, where they essentially committed to eliminating the gender gap in tenure rights and to ensure secure land rights for all women and men, particularly the more vulnerable (UN, 2015).

The Scale of the Problem: How Many Women Are Tenure Insecure?

We do not know. The data on tenure security is scarce and inconsistent. It typically comes from surveys that cover only a portion of a country, as is the case with agricultural census or surveys of urban populations; surveys that only focus on specific segments of the populations, such as land users, landholders, landowners, or landowners with documents; and surveys that do not consistently collect all the information necessary to evaluate women's tenure security. It is therefore very difficult to aggregate the information nationally and have comparable measures across countries.

A primary challenge is that most data on tenure security is collected at the household level and by default assumes that the interests of women are subsumed in the interests of their households or that all household members share congruent interests, even though when a household dissolves—for any reason—it does matter who has formal or customary rights to the land. Furthermore, because household-level surveys typically engage the head of the household, or the most knowledgeable person in the household, they likely fail to provide accurate information about the tenure security of other adult household members who may be more insecure: elderly parents or in-laws, adult children, siblings who live in the same house, and especially wives or cohabitating partners.

Lastly, surveys rarely ask about perceived tenure security. They are more likely to ask about the ways in which people access land (i.e., Do they rent it? Do they own it? How did they acquire it?), or whether people have documents to prove their land rights. All important information, of course, but not sufficient to determine whether people actually feel tenure secure and it is these perceptions that will fundamentally shape what people want to do with the land: how they want to use it, whether they want to invest on it, and whether they want to transact it, which are precisely the environmental, economic and social outcomes that policymakers want to influence.

Still, the limited evidence to which we have access points to systemic differences in women's and men's tenure security. The pioneering effort of Doss et al. (2014) to gather individual data on ownership of agricultural parcels in Ghana, Ecuador, and the state of Karnataka in India provides a preliminary glimpse into gender differences in tenure security (Fig. 5.1). In Karnataka, 72% of the men owned agricultural parcels compared to a mere 16% of the women. The gap was smaller but still substantial in Ghana, where 67% of men and only 32% of the women owned land. By strong contrast, in Ecuador, where there is considerable joint ownership of land, women were slightly more likely than men to own an agricultural parcel. Similarly, 2010 data from the FAO Gender and Land Rights Database reveals that, in most regions of the world, the vast majority of those managing or controlling agricultural holdings are men and that holdings managed by women tend to be smaller in size (Fig. 5.2) (FAO, 2011).

These figures align with what has been revealed by smaller-scale studies on differences between men's and women's ownership or management of land, but they fall short of revealing the true scale of women's tenure insecurity. They do not address the fact that even those who own or manage land may not be tenure secure or that people who access land through arrangements such as customary systems or leasing markets may in fact feel they have secure tenure.

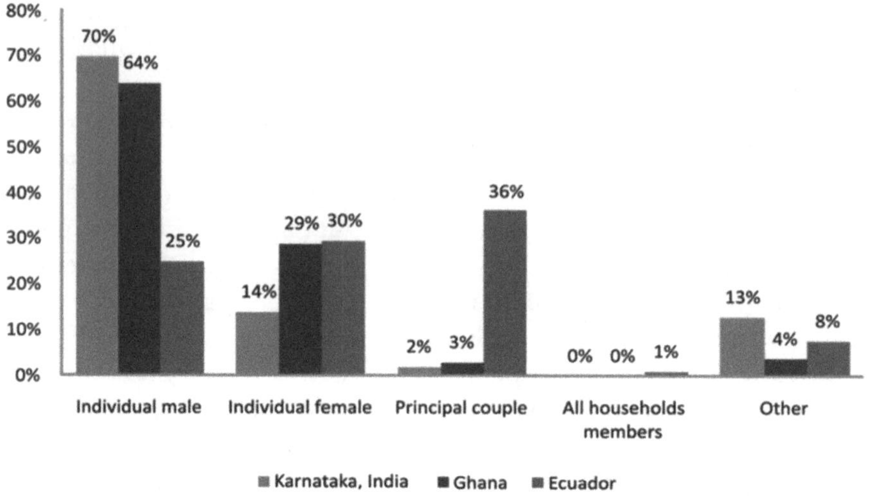

Fig. 5.1 Distribution of the form of ownership, agricultural parcels. Source: Doss et al. (2014). Reproduced with permission

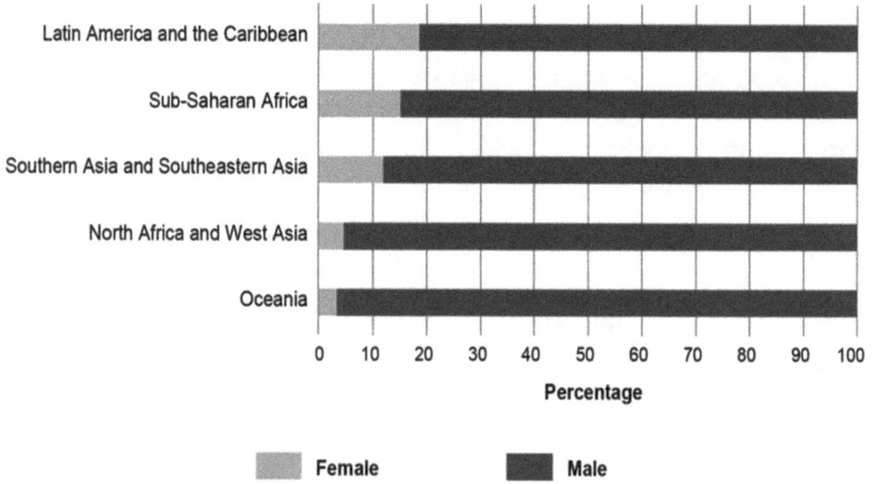

Fig. 5.2 Share of male and female agricultural holders in main developing regions. Source: Food and Agriculture Organization of the United Nations (2011), http://www.fao.org/3/i2050e/i2050e.pdf. Reproduced with permission

Fortunately, the SDGs have opened a highly consequential path to fill this data void. SDG indicators 1.4.2 and 5.a.1 will rely on primary, nationally representative, and sex-disaggregated data to diagnose and monitor changes in tenure security for all—men and women, with a particular focus on the more vulnerable among them. SDG indicator 5.a.2 will track legal and policy steps countries have taken to reduce the gender gap in tenure rights.

This is undoubtedly a tremendous breakthrough—a global agreement on the type of data to collect, national mandates to collect that data, and powerful and highly visible mechanisms to monitor progress in tenure security. In practice, however, it will take years before these commitments yield meaningful, inclusive, and up-to-date data. During this transition, Prindex, a recently launched global poll on property rights, offers some high-level findings.

For the 33 countries in which Prindex has been piloted, we now have nationally representative data on adults' perceptions of their tenure security at the individual level (Fig. 5.3).[1] The Prindex data offer a significant step forward because it asks about perceptions of tenure security; it uses questions that are consistent across countries; and most importantly, it relies on nationally representative samples of adults, not households.

In line with the SDG guidelines, Prindex gauged people's tenure security by asking them how likely they were to lose their house against their will in the next five years (Fig. 5.3). Of the nearly 53,000 adults interviewed, one in four felt insecure about their land and property (Prindex, 2019a). This represents an astonishing number of people, given that the countries included in the study have a combined population of 889 million adults.

However, Prindex data illuminated significant differences in tenure security by gender, showing that the gender gap in tenure security was unexpectedly small and inconsistent across countries (Fig. 5.4) (Prindex, 2019b). In some countries, like Benin, the UK, and Peru, women were 5% more likely than men to report being insecure in their tenure. In others, like Vietnam, Mozambique, and Malawi, no gender gap in tenure

[1] For more information on Prindex and its methodology, see www.Prindex.net.

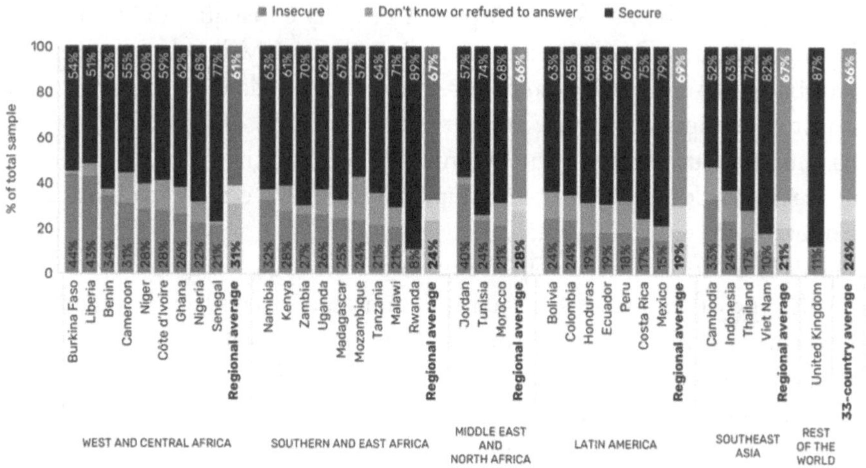

Fig. 5.3 Tenure insecurity and security for all adults, by country and region. Source: Prindex (2019a). Reproduced with permission

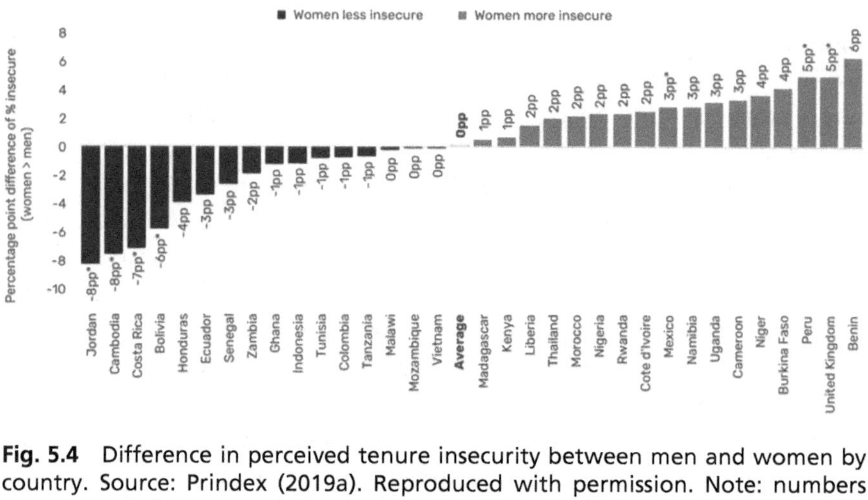

Fig. 5.4 Difference in perceived tenure insecurity between men and women by country. Source: Prindex (2019a). Reproduced with permission. Note: numbers were rounded to zero decimal places; there are small differences between countries that may be observed by the size of the bars even though the number is the same

security was found. Even more surprisingly, data from Prindex demonstrated countries where men were more likely than women to report being insecure in their tenure, such as in Bolivia, Cambodia, and Jordan.

Importantly, Prindex found that, in all but a handful of countries, women were more likely than men to say that they could lose the land in the event of divorce (Fig. 5.5) or spousal death (Fig. 5.6) (Prindex, 2019b). The differences are staggering: women are up to 46% and 35%

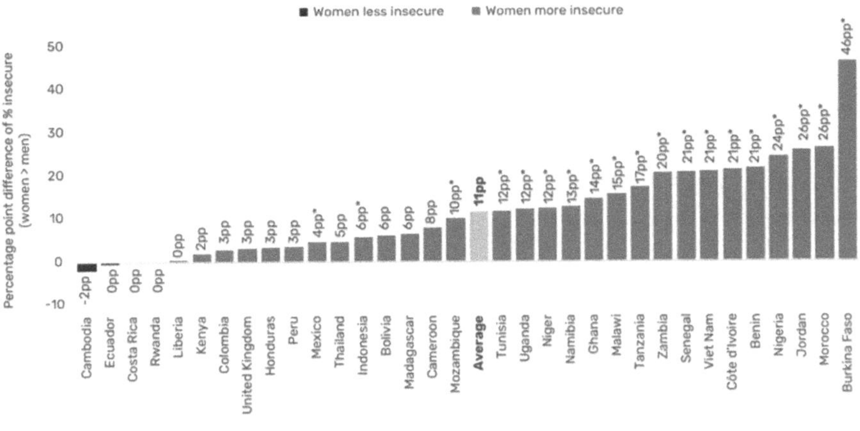

Fig. 5.5 Difference in tenure insecurity rates by gender in a divorce scenario. Source: Prindex (2019b). Reproduced with permission

Fig. 5.6 Difference in tenure insecurity rates by gender in a spousal death scenario. Source: Prindex (2019b). Reproduced with permission

more likely to feel tenure insecure from divorce and the death of a spouse, respectively. This is not trivial since, across the 33 countries, disagreements with family or relatives were the second most common reason given by those who reported tenure insecurity (Prindex, 2019a).

The Prindex data represent just 33 countries, but the data so far provide valuable insights. First, the gender disparity in tenure insecurity is indeed a critical concern. Second, overall nearly a quarter of women are tenure insecure, and the underlying reasons vary across countries. Third, we should not make assumptions about the magnitude, or even direction, of the gender gap in tenure security: each country will have to rely on its own sex-disaggregated, nationally representative data on adults' tenure security for policy decisions. Lastly, to adequately address tenure security concerns, we must identify the drivers of the insecurity. The data so far indicate a need to depart drastically from the prevalent and more traditional approaches to secure tenure, which tend to focus on (gender-blind) formalization and mapping of property rights, decentralization of government responsibilities, and digitalization of land administration services.

To enhance women's tenure security, the suite of interventions needed must also include efforts to protect women from discriminatory cultural norms and family dynamics seeking, for example, to achieve community-wide changes in attitudes and behaviors that enable daughters to inherit rights to land and protect women from losing their land rights when they get married, get divorced or become widows; or gender-responsible government programs so that women are not consistently left behind when land rights are allocated, adjudicated, or compensated for, when information about laws, programs or opportunities is disseminated, or when decision-making authority is devolved to community members.

What Can Be Done to Address Women's Tenure Insecurity?

There are no simple cookie-cutter solutions that apply across the world. It is important to consider (1) whose tenure security we hope to strengthen, and specifically which group of women; (2) how the insecurity manifests itself, or put differently, what more specific security constraints we are trying to address; and (3) how those constraints can be addressed effectively. We address each of these questions below.

Who Are the Tenure Insecure Women?

Not all women are tenure insecure, as is clear from the data presented above. We therefore need a more nuanced understanding of who the tenure insecure women are, and our findings are bound to vary by country and context (Chigbu et al., 2019).

Consider, for instance, the factors that might drive differences in the tenure security of women across the world. Are tenure insecure women concentrated in rural areas or in informal settlements? Are they more prevalent among certain ethnic or religious groups, in indigenous communities, or among the elderly? Are they concentrated in low-income populations or among small business owners? Given the significance of marital status, are most tenure insecure women widows, common-law partners, polygamous wives, unmarried women? Answers to these questions will shed light on who needs to be targeted and how best to reach them to strengthen their tenure security (e.g., through printed materials, radio shows, text messages, self-help groups, schools, religious institutions).

Why Are Women Tenure Insecure?

Just as there is variation in which women have insecure tenure, there is also variation in the causes of tenure insecurity. Some women may be affected by tenure security constraints that impact not just them, but their entire community. This may be the case of women who are members of indigenous communities with insecure land rights or women who belong to communities subjected to poor governance at the village, district, or provincial level. Other women experience tenure insecurity because their households are insecure. When families are dealing with boundary disputes, or are unable to pay costly fees, or are discriminated against for political reasons, the challenges they face will inevitably impact their female members. And yet another group of women experience tenure insecurity even though (men in) their families and their communities have secure tenure.

Identifying a suite of interventions that can successfully strengthen women's tenure security requires that one starts by fully identifying the problems. What are women's greatest concerns? Research indicates that concerns can range from their families being displaced by corporations, their in-laws taking over the land, their inability to pay rent, the threat of customary leaders reassigning land use rights and cultural norms dictating that women should forego their rights to inherit parental land in favor of their brothers, to women being ignored by authorities when there is a need to resolve land disputes. The main sources of women's tenure insecurity can be revealed using a combination of carefully tailored surveys, qualitative research, and input from gender experts on land.

But one should go further to determine the root of these insecurities. What women experience is the result of interrelated forces that shape land tenure and rights in their particular context—what is typically known as a "land rights system." To identify the weaknesses in a land rights system, one should consider four of its elements. Land rights systems are shaped by:

- Statutory and customary land-related laws, policies, regulations, conventions, and agreements that embody the rights determined and enforced by governments and communities;
- Formal and informal institutions and actors who influence, decide, manage, or enforce land-related rights;
- Social norms that shape attitudes and beliefs on who should have land, for what purpose, and through which means; and,
- Individuals and communities whose land-related rights are protected, strengthened, limited, or negated by the system.

Challenges to women's tenure security can originate or be reinforced by any one of these elements. For women to be tenure secure, as laid out in Landesa (2019), the land rights system must be effective, inclusive, and gender-equitable.

Women will not be secure, unless the land rights system in which they operate works. The land rights system has to be able to clearly define and enforce who has which rights and for how long, as well as how people can acquire new rights and how people can protect the rights they possess (Landesa, 2019; Doss & Meinzen-Dick, 2018; Place et al., 1994).

However, land rights systems can operate effectively and still fail to include and protect some members of the society. Thus, for women to be tenure secure, the land rights system has to ensure land rights for women of all ethnicity, religion, or other social groups, and in line with their needs and perspectives (Landesa, 2019).

Furthermore, to ensure women's tenure security, the land rights system must also rule out gender preferences in how land rights are acquired, experienced, and protected. Meeting this higher bar requires paying attention to a much broader set of policies, laws, norms, institutions, and actors, including, for example, communities' land use plans, and religious, family, inheritance, and marital laws that also shape and enforce women's ability to acquire, exercise and protect their land rights.

What Can Be Done?

As the saying goes, identifying the problem is halfway to the solution. We cannot provide an exhaustive list of options here, as that is well beyond the scope and focus of this chapter. However, a few recommendations may prove helpful.

First, not all instances of tenure insecurity require deliberately targeting women. Even when interventions seek to strengthen tenure security at the broader community or household level, they should be designed and implemented using gender-responsive, rather than gender-blind, goals and practices. Only then will efforts to secure tenure for many avoid unintentionally leaving women behind.

Further, while information about women's perceived tenure insecurity is critical to guide interventions, women's perceptions will not capture the impact of tenure threats unknown to them. This is of particular relevance in environments where information is unevenly shared and received, and, where marginalized populations and, women in particular, may not be privy to either legal regulations or to plans by the government, corporations, or their local chiefs. Because of these constraints, women may incorrectly believe they possess more tenure security than they actually have. Efforts to strengthen women's tenure security should therefore rely on experts and key stakeholders to gather and integrate this additional information.

Second, since the challenges to be addressed are likely to include deeply rooted gender-discriminatory attitudes, behaviors, and systems that shape social norms and permeate governance structures, the solutions need not be restricted to what has been tried within the land rights sector. We can learn, for example, from what has been done around the world to contain the HIV-AIDS epidemic or to address gender-based violence, child marriage, female genital mutilation, and other challenges that are also complicated and exacerbated by systemic discrimination.[2]

Innovative and effective options need not be limited to those originating in the development sector. There is much that can be learned, for instance, by the way private companies have mastered their ability to deliver information, target messages, provide low-cost services, and shape our preferences and behavior. Sodas, cell phones, jeans, social media, and animated characters have reached people in every corner of the world in astonishingly short periods of time. This creativity is integral to breaking through structural and systemic biases that can be extremely resistant to change.

Lastly, efforts to enhance women's tenure security are more likely to succeed if they take a holistic approach. This means combining a number of interventions. Interventions may be top-down, bottom-up, peer-to-peer; delivered through multiple entry points (e.g., broad-reaching media, authorities, word of mouth, or using demonstration effects); and targeting different stakeholders (e.g., women, men, the elderly, the youth, and authorities) with complementary goals; and taken together, these approaches can offer a combination of quick gains and sustained change.

This is virtually impossible to accomplish by one set of actors. Policymakers, government agencies, community leaders, and civil society organizations each have mandates, resources, and capacity constraints. Thus, champions of women's tenure security will do well by promoting, from the beginning,

[2] The Population Council, the CDC, and other organizations and agencies have done extensive research that is publicly available and easy to access.

approaches that rely on multipronged interventions and a sustainable coalition of changemakers. A recent example from Landesa's work follows.

A Multipronged Approach to Strengthen Women's Land Rights in Uttar Pradesh, India

To ensure women in Uttar Pradesh are better placed to acquire land rights through family inheritance or government land allocation programs, Landesa has partnered with government officials, local representatives, civil society organizations, lawyers, and the media. Over the years, we have operated at several nodes:

- *We have imparted land literacy trainings to women to enhance their understanding of their land rights and about how to navigate the land administration system.*
- *We have sensitized local Land Revenue officials so that they could better understand the regulations that affect women's land rights and could apply them confidently.*
- *We have sensitized local elected representatives who are members of the village-level Land Management Committees so that they could better determine who is eligible to receive land parcels.*
- *We have collaborated with the Department of Panchayati Raj to ensure that the curriculum to train newly elected representatives included information on women's land rights and the processes to follow in allocating land to them.*
- *We have engaged with District Administrative Offices to get their buy-in and enlist their support so that laws were implemented effectively and a range of programs acted in coordination. This resulted, for example, in the Rural Development Department writing the legal provisions supporting women's land rights on their office walls to make them highly visibly, and coordinating with us so that our literacy sessions were carried out in the same areas where they were empowering women collectives.*
- *We have engaged with the Uttar Pradesh Revenue Department on an ongoing basis, asking them to support the work and advocating for changes in legal entitlements and procedural laws that were constraining women's land rights.*
- *We partnered with civil society organizations in Uttar Pradesh so that they could better understand the intricacies of women's land rights and could advocate with us for the changes needed.*
- *We partnered with lawyers to gain a more in-depth and nuanced understanding of the practical implications and challenges around women's land rights.*
- *We worked with the media to elevate the challenges of women's land rights and gradually build broad-based fruitful conversations to address these challenges.*

Conclusions

While much has been done and learned around women's land rights over the past few decades, securing tenure rights for all women and men by 2030, as promised under the SDGs that tackle global poverty, hunger, and gender inequality, requires addressing the critical gaps that remain. We need to ensure that there are simple and practical ways to generate, access, and use primary data to diagnose and track progress on women's land rights. We need to ensure that there are nuanced and contextual assessments of the key sources of tenure insecurity for women. We need carefully assessed and documented examples of what works (and what has not worked), when and for whom so that women's land rights can be addressed in an effective and timely manner. We, practitioners, governments, and funders, need to acknowledge that these issues are complex, that addressing them requires a longer time horizon and complementary interventions, and, as put forth by the Bridge Collaborative (www.bridgecollaborativeglobal.org), that it necessitates bold approaches that benefit from multiple disciplines and expertise from the start.

References

Allendorf, K. (2007). Do women's land rights promote empowerment and child health in Nepal? *World Development, 35*(11), 1975–1988.

Archambault, C., & Zoomers, A. (2015). *Global trends in land tenure reform. Gender impacts*. Routledge.

Benschop, M. (2004). *Women's rights to land and property*. Women in Human Settlements Development—Challenges and Opportunities. UN-Habitat submission to the Commission on Sustainable Development.

Chigbu, U. E., Paradza, G., & Dachaga, W. (2019). Differentiations in women's land tenure experiences: 21-implications for women's land access and tenure security in Sub-Saharan Africa. *Land, 2019*(8), 2.

Deininger, K., Ali, D. A., & Yamano, T. (2008). Legal knowledge and economic development: The case of land rights in Uganda. *Land Economics, 84*(4), 593–619.

Dillon, B., & Voena, A. (2018). Widows' land rights and agricultural investment. *Journal of Development Economics, 135*(C), 449–460.
Doss, C., Deere, C. D., Oduro, A. D., & Swaminathan, H. (2014). The gender asset and wealth gaps. *Development, 57*(3–4), 400–409.
Doss, C., & Meinzen-Dick, R. (2018). *Women's land tenure security: A conceptual framework*. Research Consortium. https://consortium.resourceequity.org/conceptual-framework
Ghebru, H., & Holden, S. (2013). Links between tenure security and food security: Evidence from Ethiopia. *IFPRI Discussion Papers* 1288. International Food Policy Research Institute (IFPRI).
Goldman, M., Davis, A., & Little, J. (2016). Controlling land they call their own: Access and women's empowerment in Northern Tanzania. *Journal of Peasant Studies, 43*(4), 777–797.
Grabe, S. (2015). Participation: Structural and relational power and Maasai women's political subjectivity in Tanzania. *Feminism & Psychology, 25*(4), 528–548.
Holden, S., Deininger, K., & Ghebru, H. (2011). Tenure insecurity, gender, low-cost land certification and land rental market participation in Ethiopia. *Journal of Development Studies, 47*(1), 31–47.
Landesa. (2019). *Towards a gender-responsive land rights system*. Landesa.
Lastarria-Cornhiel, S., Berhman, J. A., Meinzen-Dick, R., & Quisumbing, A. (2014). Gender equity and land: Toward secure and effective access for rural women. In A. Quisumbing, R. Meinzen-Dick, T. Raney, A. Croppenstedt, J. A. Berhman, & A. Peterman (Eds.), *Gender in agriculture. Closing the knowledge gap*. Food and Agriculture Organization.
Meinzen-Dick, R., Quisumbing, A., Doss, C., & Theis, S. (2019). Women's land rights as a pathway to poverty reduction: A framework and review of available evidence. *Agricultural Systems, 172*, 72–82.
OHCHR. (2012). *Women and the right to adequate housing*. Office of the High Commissioner United Nations Human Rights.
Place, F., Roth, M., & Hazell, P. (1994). Land tenure security and agricultural performance in Africa: Overview of research methodology. In J. W. Bruce & S. Migot-Adholla (Eds.), *Searching for Land Tenure Security in Africa*. World Bank.
Prindex. (2019a). *Comparative report*. Prindex.
Prindex. (2019b). *Women's perceptions of tenure security. Evidence from 33 countries*. Prindex.

Santos, F., Fletschner, D., Savath, V., & Peterman, A. (2014). Can government-allocated land contribute to food security? Intrahousehold analysis of West Bengal's microplot allocation program. *World Development, 64*, 860–872.
United Nations. (2015). *Resolution adopted by the general assembly on 25 September 2015.*
United Nations, Food and Agriculture Organization. (2002). *Land tenure and rural development* (Land Tenure Studies #3). Food and Agriculture Organization.
United Nations, Food and Agriculture Organization. (2011). Women in agriculture: Closing the gender gap for development. In *The state of food and agriculture 2010–2011*. Food and Agriculture Organization.
World Bank. (2008). *Agriculture for development.* World Development Report 2008, A. de Janvry & D. Byerlee (Eds.). World Bank.

6

People-Land Relationships on the Path to Sustainable Food Security

Malcolm Childress, Pranab Choudhury, and Jolyne Sanjak

The 1996 World Food Summit defined food security stating, "Food security exists when all people, at all times, have physical and economic access to sufficient, safe and nutritious food that meets their dietary needs and food preferences for an active and healthy lifestyle." This definition is still commonly used. Accordingly, food security programming and policies consider needs within three underlying pillars and one cross-cutting pillar: food availability, food access, food utilization (nutrition), and the stability of these three over time.

M. Childress (✉)
Global Land Alliance, Washington, DC, USA
e-mail: mchildress1@landallianceinc.org

P. Choudhury
NRMC-Center for Land Governance, Bhubaneswar, Odisha, India

J. Sanjak
Tetra Tech, Washington, DC, USA

M. B. Holland et al. (eds.), *Land Tenure Security and Sustainable Development*,
https://doi.org/10.1007/978-3-030-81881-4_6

The pillars of food security are intimately tied to the land, and problems related to food security and to sustainable agriculture are complex and interconnected with the land. People-land relationships (PLR) comprise the totality of the human-land interaction including current and historical use and occupancy, concepts about the meaning and value of land, and include a subset of relationships concerning the nature and quality of land rights, tenure forms, and related institutions which jointly generate and condition land tenure security. We refer to this subset of relationships as land tenure security (LTS). PLR affects the *availability*, *access*, and *utilization* of food (FAO). This is because these relationships, and LTS in particular, influence the way women, men, groups, and entities access, control, and use the land; how they allocate land to different uses and among different users; how they can access inputs, extension services, and entitlements, and the types of investments they make in land productivity and conservation. LTS also interacts with natural or man-made shocks and hazards, such as those induced by climate change, civil conflict, and demographic shifts, in ways that affect the *stability* of availability, access, and utilization of food.

It is, therefore, unsurprising that food security is so intimately tied to PLR and shaped by LTS. People have always depended on the land for basic needs (e.g., food, water, shelter, and defense) and for emerging wants (e.g., materials for producing things, places to enjoy, and food beyond subsistence). Since the dawn of human society, people have had rules about who can access land for what purposes, over what timeframe, and under what conditions (i.e., from the territorial understandings of hunter-gatherer groups to the most recent legal constructions of development and food security as human rights). These rules are dynamic, changing as the need for, and the nature of, rules evolve in accordance with changes in people's relationship to nature and in nature itself. These rules also shape human behavior over how land is managed, accessed, and utilized, influencing whether and how land will be sustainably managed, and food sustainably produced. Yet, with over 50 years of attention in global development policy, theory, research, and programs addressing LTS and food security, too little attention is paid to the interconnected ways in which the different aspects of LTS combine to influence food security. Understanding these pathways has implications for a more holistic approach to LTS and improving food security across communities of practice in agriculture, environment, and governance.

Today, food security remains an aspiration, and tenure security is elusive for many people. Historical, political, and demographic factors have inhibited progress toward securing and ensuring food security (see, e.g., Maxwell & Wiebe, 1998, p. 31, for example). The global development discourse increasingly links PLR with the widespread adoption of sustainable agriculture and improvements in food security, and leaders are making more significant commitments to addressing LTS constraints. Yet, attention to unique LTS constraints and integrative programming addressing these constraints are still inadequate compared to the scope of challenges.

In this chapter, we describe the evolving landscape of the global development discourse around LTS and food security by synthesizing diverse strands of literature to build a heuristic case for integrative solutions for strengthening LTS to achieve sustainable food security. We then present evidence about the status of these linkages, comprising five dimensions of the LTS–food security relationship: (1) aggregate land scarcity, (2) land access and inequality, (3) incentives and risks, (4) gender, and (5) shocks and hazards. These dimensions are then mapped on to two key policy agendas for food security: Sustainable Development Goal 2 and the FAO definition of food security. To bring the need for integrated action alive, we briefly characterize how LTS elements are linked to food security patterns using a dataset from South Asia.

LTS and Sustainable Food Security in the Global Development Rhetoric

LTS features prominently in international development program discourse and is increasingly featured in international agreements. Improving LTS or the quality of property rights is generally recognized as an enabling condition for improving food security. Differences across programs and stakeholders are typically about the people or entities targeted for LTS improvement, the level of aggregation (individual, household, community, region, nation, etc.), the form of tenure or rights to prioritize, and the institutions, methods, or tools to strengthen LTS.

The scope of the LTS challenge itself is only beginning to be understood at a global level as comparable metrics become available. The Prindex survey of citizens' perceptions of LTS in 33 countries in 2018, for example, showed that one out of four adults perceived there was a risk of losing their home or other property, including agricultural land used for growing and selling food, within five years.

Although the evidence on the relationship between LTS and food security is still emerging, the theory of change motivating why LTS will improve food security can be found within a broad range of bilateral and multilateral development assistance documents. Strengthening LTS is assumed to lead to food security through two main pathways: (1) through creating greater certainty for those living on the land to make investments that can increase food production and (2) through improving income generation, thus enabling people to buy more food or to have access to services and markets that would otherwise be difficult to reach, such as agricultural extension and credit in some contexts.

Over the last 10 years, documents framing food security programming increasingly include explicit links to LTS, which has led to increased investments in strengthening LTS. For example, commitments from the 2009 G8 Summit to achieve global food security led to the US-launched Feed the Future Initiative. Feed the Future's programmatic strategy and results from framework incorporated LTS within an agricultural policy agenda (Lawson et al., 2013). At around the same time, the FAO's Land Tenure Service engaged stakeholders to introduce global good practice guidelines on land governance. Ultimately adopted by the FAO Committee on World Food Security in 2012, The Voluntary Guidelines on the Responsible Governance of the Tenure of Land, Fisheries, and Forests (VGGTs) (Box 6.1) gave even more visibility to the connections of LTS and food security and reflected a new consensus on how to achieve LTS.

Box 6.1 The VGGT

The purpose of these Voluntary Guidelines is to serve as a reference and to provide guidance for improving "*the governance of tenure of land, fisheries and forests with the overarching goal of achieving food security for all and to support the progressive realization of the right to adequate food in the context of national food security*" (FAO and the Post-2015 Development Agenda–Tenure Rights).

In addition, an uptick in support for integrating LTS measures or activities in food security programming has occurred since 2012 (e.g., see recent USAID training material). VGGT-specific activities and VGGT influence on the design of LTS measures, activities, and policies also occurred. Since 2012, the FAO has directly supported activities in 31 countries, as well as regional activities in the Sahel and Mekong region leading to legal and/or policy changes. Yet, progress still lags in comparison to the scope of the challenges, and the VGGT has not achieved widespread uptake as a voluntary "soft-law" instrument as envisaged, nor has it been championed by multilateral institutions beyond the FAO (Via Campesina 2015).

In tandem with these activities, LTS has gained a presence in agendas and commitments of regional and global convenings of intergovernmental bodies within the past 20 years. Some notable examples, in addition to the Committee on World Food Security (CFS), include the 2001 Summit of the Americas, the 2002 UN World Summit on Sustainable Development, 2008 UN CSD-16, the 2009 13th Ordinary Session of the African Union Assembly of Heads of State and Government Declaration on Land Issues and Challenges in Africa, the 2013 G8 Summit Partnerships on Land Governance, the 2015 adoption of the 2030 Agenda for Sustainable Development, and the 2018 Commission on the Status of Women. These recognize the need to strengthen LTS for general development outcomes, as well as for food security.

Perhaps most importantly for global development rhetoric, LTS is at the heart of many of the Sustainable Development Goals (SDGs). LTS is explicitly referenced in Goal 1 (No Poverty), Target 1.4; Goal 2 (Zero Hunger), Target 2.3; and Goal 5 (Gender Equality), Targets 5.A.1 and 5.A.2. With these commitments, it was both strategic and welcome to the land community that the UN Commission on the Status of Women in both 2016 and 2018 included language on women's land rights, aligning with a long-standing interest in the topic by participating civil society organizations.

Adoption of sustainable agricultural practices is important for many of the SDGs, but it is at the heart of Goal 2 which commits to ending hunger, achieving food security, improving nutrition, and promoting sustainable agriculture. Sustainable agricultural practices often require considerable investments, and farmers, in general, might not realize gains for years. As we discuss below, LTS and tenure forms are a major component of land investment decisions. While LTS is only mentioned explicitly under Target 2.3, among the eight targets set for Goal 2 LTS is relevant for Target 2.1 on universal access to food; Target 2.2 on ending malnutrition; Target 2.3 on doubling productivity and income for small-scale food producers; and Target 2.4 on sustainable food production and resilient practices.

Storylines and Evidence Emerge from a Look at the Literature

The land is the natural resource base for food production. Food security and the long-term sustainability of the world food system are issues of major concern (FAO, 2019, IPCC 2019). The current food system supplies food successfully to much of the world's population; however, an estimated 821 million people are currently undernourished, hundreds of millions more face occasional hunger, and hundreds of millions more live on the edge of hunger.

Despite the importance of land—and thus LTS—for food security, surprisingly few studies assess how changes in LTS directly impact food security or have effects on drivers of food security via changes in economic variables (e.g., household income). Few studies estimate the direct impact of improved LTS on direct measures of food security, such as the Food Insecurity Experience Scale. Within this limited literature, Maxwell and Wiebe (1999) and Holden and Ghebru (2016) offer comprehensive logic models that connect LTS elements with food security and provide evidence for their main tenets. Simply summarized, these include (1) that LTS drives positive changes in investment, access to credit, and land transfers, which in turn improve production efficiency that improves food security, and (2) that property rights directly affect food security through land access and indirectly affect it by influencing land transfers and thus production efficiency.

Nkomoki (2018), Baltissen and Betsema (2016), and Espinosa (2014) provide Africa-wide literature reviews examining the relationship between forms of tenure, sustainable agricultural practices, and food security. Nkomoki (2018), for example, presents evidence in Zambia that there is a lower probability of farmers adopting crop diversification, agroforestry, and planting basins when LTS is weaker, thus increasing risks to food security. More generally, as pointed out in Maxwell and Wiebe (1999) and Holden and Ghebru (2016), a primary barrier to achieving a greater understanding of the LTS–food security relationship is that research and discourse on these topics are generally siloed within narrow, often project-driven, thematic categories.

While these authors appropriately call for more research to garner direct evidence for how LTS affects food security, they also share our viewpoint that diverse but distinct strands of research support the hypothesis that improvements in LTS are linked to food security. Baltissen and Betsema (2016) observe that, "Much information exists on the links between land governance and food security in Africa, including academic research, policy reports and case studies. It is, however, not always clear where to find this information."

Table 6.1 maps the dimensions of LTS to the pillars of food security. In particular, the table demonstrates how each pillar of food security is intimately tied to various dimensions of LTS that go beyond the effects of LTS on increasing parcel-level investments (an area of much intense focus on the literature) and articulates other links, such as the effects on the stability of food supply from wider scale impacts of land institutions.

In the following sections, we try to open up the silos to present a narrative about the most direct and significant ways that LTS matters for food security. Our aim is to organize findings from across a large, disciplinarily dispersed and context-varied spectrum of the qualitative and quantitative literature. To do so, we categorize the LTS–food security relationship into five dimensions:

1. Aggregate land scarcity
2. Land access and inequality
3. Incentives and risks
4. Gender
5. Shocks and hazards

Table 6.1 Dimensions of LTS mapped to pillars of food security (FAO)

	Food availability	Food access	Food utilization	Stability of food supply and access[a]
Land tenure security	Propensity to invest in land to increase productivity, such as soil conservation and water management Reduced loss (technology adoption and conflict resolution)	Potential gender, age, and ethnic differences related to the intrahousehold and intracommunity distribution of access to land, production, and income generation		Migration patterns Climate change Civil conflict
Tenure forms **Property rights** **Institutions**	Land usage Efficiency of allocation	Macro-growth (quality property rights systems and with less inequality in the assignment of rights) Household access to land (women's garden plot allocation; land rental and purchase markets, land redistribution)	Nutrition and health outcomes (particularly, women's LTS)	Land takings; expropriation without compensation or due process; forced expulsion by individuals or groups; uninformed or irresponsible large-scale private investment

[a]The relationships, particularly in this column, might be bidirectional in terms of causality with LTS issues affecting and being affected by the factors listed; there can be a vicious cycle of food and land insecurities

Aggregate Land Scarcity Exacerbated by Population Growth, Unsustainable Land Use, and Climate Change

This narrative starts with the land resource base itself. The overall quantity of land available for food production has become a major issue of concern, as land suitable for allocation for food production is scarce, and much of what remains is degrading. The ecosystem services—including supply of water and soil formation—that are critical inputs for food production are also under threat. According to the most recent Intergovernmental Panel on Climate Change (IPCC) (P. R, Shukla, 2019), human use currently affects more than 70% of the land on earth. Humans currently use one quarter to one-third of land's potential net primary production for food, feed, fiber, timber, and energy. About a quarter of the Earth's ice-free land area is subject to human-induced degradation. Soil erosion from agricultural fields is much higher than the soil formation rate, especially under conventional tillage. Climate change exacerbates land degradation, particularly in low-lying coastal areas, river deltas, drylands, and permafrost areas. Drylands affected by drought are increasing. In 2015, about 500 million people lived within areas that experienced desertification between the 1980s and 2000s.

As human use of land reaches natural limits, the allocation, use, and management of the land under current use become critical topics for food security. As populations increase their income and change their food preferences, the food system and land face pressures to intensify as, for instance, demand for animal-sourced products increases. A 50-year time series from IPCC shows that population growth and changes in per capita consumption of food, feed, fiber, timber, and energy have led to high rates of expansion of land and freshwater use, with agriculture currently accounting for approximately 70% of freshwater use. Other population and demographic dynamics can also perpetuate land use choices that exacerbate scarcity of land for food production, such as urban expansion on to prime agricultural land, or migration on to already fragile lands due to poverty and conflict dynamics.

Expansion of agriculture and forestry, supported by increases in productivity-enhanced agriculture, has enabled consumption and food availability for a growing global population, but this expansion may soon run into natural limits. With large regional variation, these changes have contributed to increasing net greenhouse gas emissions and the loss of natural ecosystems (e.g., forests, savannahs, natural grasslands, and wetlands) and biodiversity. In this situation, policies affecting LTS have a major role in managing the trade-offs of land use for agricultural expansion and intensification, timber, energy, and conservation of ecosystem services to supply food systems with their required natural inputs.

Land Access and Inequality

Access to land (e.g., through land reallocation or allocation policies) supports food self-sufficiency (Holden & Ghebru, 2016). For many households living in areas with limited labor market opportunities, household food production on household-controlled land is necessary as a complement to food purchases. In many cases, household food production is the only source of food and nutrition. Insufficient access to land (including landlessness), inequality in land distribution, and the lack of means to access land can be significant threats. Muraoka and Jayne (2018) established a strong linkage between food security and land access in Kenya and found that renting-in land helps with household-level food security. They also showed that long-term productivity and investment are lower on rented plots, indicating that households on rented lands do not fully realize the potential of the land to contribute to food security even if it provides household food security in the near term. In another study, Keswell and Carter (2014) showed significant positive impacts on household well-being, including food consumption, from land access gained through South Africa's land reform program.

Even with sufficient land access, food security is threatened when households do not have the income to purchase food. Many people living in poverty in rural or peri-urban areas still depend on land access for their livelihoods. The observed livelihood strategies of many households at risk of food insecurity are to diversify into multiple land-based and non-land-based activities (migration of some members, seasonal migration, on and off-farm work; Paudel et al., 2017), often with land-based activities forming the core of nutritional subsistence.

Concerns over access to land and inequality in its distribution are heightened in the context of expanding large-scale agriculture and other types of large-scale land-based investments (LSLBI). While it has been generally agreed that significant investment in agriculture is needed to feed the world's growing population, the way in which these investments happen can present major challenges for vulnerable groups. One challenge occurs when large-scale agriculture expands through capital-intensive pathways and displaces previous agricultural laborers or smallholders, which may worsen food security for the displaced population if other labor opportunities do not materialize (Hufe & Heuermann, 2017).

A surge in LSLBI drew attention almost a decade ago to the reality that not all LSLBI are carried out responsibly. All too often, even when intentions are to benefit local populations, property rights of local people are not recognized or respected and LTS worsens. Deininger and Byerlee (2011) and many other authors have described such patterns and impacts over the last decade. In 2014, the FAO CFS endorsed Principles of Responsible Investment in Agriculture and Food Systems (RAI) to promote improved investment practices and encourage business models that benefit local smallholder farmers or other vulnerable people. Adherence to RAI remains limited, however, like the VGGT.

The relationship of overall inequality in landownership with depressed economic growth and lower levels of poverty reduction has become increasingly clear with recent analyses and meta-analyses (Cipollina et al., 2018; Deininger, 2003; Deininger et al., 2009; Deininger & Squire, 1998; FAO, et al., 2019; Fort, 2007). Unequal landholding and reduced growth result in greater levels of food insecurity than would otherwise be the case (Box 6.2). Although large-scale redistributive land reform has receded from the forefront of development policy in the twenty-first century, these relationships should make policymakers wary of policy directions that further concentrate landownership. Further, the evidence so far

Box 6.2 A Perspective on Asset Inequality and Food Insecurity

"The greater the in asset distribution such as land, water, ..., the more difficult it is for the poor to participate in economic growth processes. This then slows the progress in reducing food insecurity and malnutrition. ... Land-resource scarcity and inequities are growing, with poor and marginalized population groups worldwide often having the least access to land. They are confined to 'poverty traps' of marginal and degraded lands ... vulnerable to climate variability and have no secure tenure." (FAO et al., 2019).

indicates caution about investments that can disenfranchise the poor of their existing property rights to land and should catalyze consideration of approaches that improve access to land (e.g., to improve the flexibility and inclusiveness of land markets or by allocation of public lands) to increase food security.

Incentives and Risks

Because land is a critical and productive asset, LTS can affect food security through incentives and risks perceived by households, particularly in the context of rural communities in low and middle-income countries. The quality of household-level LTS influences investment choices and asset values, and thus food security and adoption of sustainable agricultural practices. With strong LTS and enforceable and documented property rights, the household has greater certainty in obtaining the benefits from investment in, and the use or transfer of, a property in subsequent time periods (or low risk of not obtaining them). Depending on other contextual factors, the rightsholders can improve access to credit and other resources because land provides collateral (see discussion in India about public entitlements connected to property rights documentation).

Bowen and Ngeleza (2019) identify five pathways through which LTS-driven investments can increase household incomes and thus food security: land-attached investment (e.g., soil and water conservation, livestock, machinery, crops and trees, and housing) and cropping decisions; private infrastructure (e.g., wells and pumps); reduced environmental damages; land transfers; and access to, and the cost of, finance. There is ample evidence that strong LTS can increase a household's propensity to invest in productivity enhancing, durable investments, including soil conservation and practices that enhance ecosystem services, and that land values can increase due to these investments. A range of other results also confirm the strength of the five pathways, although most do not directly document income or productivity increases (IFPRI, 2019; Lisher-Witriol, 2019).

Box 6.3 Social Risk Management and Locational Value

The concept of social risk management (SRM) extends the traditional framework of social protection to include prevention, mitigation, and coping strategies to protect basic livelihoods and promote appropriate risk-taking behaviors. Social risk focuses specifically on the poor, who are the most vulnerable to, and more likely to suffer in the face of, economic shocks. Through its strategies, SRM aims to reduce the vulnerability of the poor and encourage them to participate in riskier but higher-return activities in order to transition out of chronic poverty, thereby achieve lasting food security. In addition to land's productive values, and land's values as an asset, land also has importance for SRM through its locational value.

Locational value is the increment of land value derived from its specific location in space, for example, the value derived from being located near a source of water, or a road or market or city center due to differentials in costs of production and/or transportation, or through positive spillovers of market access.

Another way to understand the importance of improving household-level LTS is to consider the role of asset security in social risk management (Box 6.3). As noted by Michael Sherraden almost 30 years ago in his book *Assets for the Poor*, assets are key factors that influence if and how people might change the way they think and behave: "Income only maintains consumption, but assets change the way people think and interact with the world. With assets, people begin to think in the long-term and pursue long-term goals" (Sherraden, 1991). The LTS attributes of assets, including the location of the asset and the portfolio of assets, are critical not only for managing the short-term risk of being expelled from a place or from using a plot but also for building pathways out of poverty over time for long-term food security.

The key linkage of LTS for food security from this perspective is that even relatively small fixed assets such as a household plot can be leveraged for productive, asset and locational values, and can play crucial roles in both short-term responses to shocks and in long-term asset accumulation. All of these dynamic consequences of LTS contribute to the stability of food security. The location and location-specific context of land can impact individual and household opportunities and outcomes because these factors affect the available assets and livelihood options, and thus a household's ability to respond to short-term shocks and build long-term investments (Jorgensen & Siegel, 2019).

The location-context of land can affect a household's food security via incentives and risks through a number of relationships. These include the land's proximity to markets and urban centers, positive spillovers of market access, productive potential, vulnerability to natural disasters, and the availability and quality of public infrastructure and public services.

With all of these relationships in play, one of the biggest determinants of incentives and risks at any point in time over the lifecycle of the household is location. The location of land assets in relation to the set of factors mentioned above is thus critical for food security. The lack of a favorable, stable location may create high food insecurity in times of instability in wages or in the case of disasters. Fadorable, stable locations may mitigate these adverse effects by allowing households to produce their own food, increase the capacity to store food or assets such as livestock, and allow greater access to labor markets and community support. Furthermore, a favorable, stable location may help individuals or households to access social programs, such as subsidies, education, credit, and infrastructure (e.g., irrigation and roads) supportive of long-term food security.

The concept of land as a key asset for SRM points to the value of policies that make land in favorable, stable locations available over the lifecycle of households. These include micro-plots and house-plot allocation, longer-term leasing or use rights, access to commons or collective areas, and attention to asset portfolios, such as land plot and livestock or land plot and fruit trees (Landesa 2020, FAO 2018).

Gender

Gender disparities in LTS are directly linked to food security. The FAO (2019), Espinosa (2014), and others explain that, at the household-level, gender differences in access to, and control of, assets like land are determined by who has decision-making power within the household. Decision-making power may be legal or be based on social and cultural norms and customs. Here, the logic is that the greater access and control of assets and resources an individual has, the greater decision-making power they will yield in a household. For instance, research on intra-household decision-making between men and women has consistently found that resources, such as incomes, are rarely completely shared or pooled between men and women within households. The power and decision-making imbalance between men and women can exacerbate women's poverty status and food insecurity during periods of economic slowdown or downturn. Further, these imbalances may make women and children especially vulnerable to changes in family status, such as the death of a spouse, as legal and social norms can mean assets, such as land, are not inherited by women.

In considering gendered aspects of how LTS and food security are interrelated, Espinosa (2014) points out that similar incentive effects discussed above have been demonstrated to result from strengthening women's land rights both within and independently of households.

There is growing and compelling literature on the impact of improving women's LTS across multiple areas linked to food security. This includes increases in investments in land and water conservation measures, increases in household spending on food and education, improvements in children's nutrition, improvements in a household's access to markets, elevated women's status in their communities, and more sustainable pathways of migration and urbanization (Box 6.4). These types of results are found across a diverse set of contexts, indicating the robustness of this dimension.

In sum, the nature of landownership and access among household members can improve or reduce food security for individual women, particularly upon inflection points of family change (i.e., marriage, divorce, and death). Given the structure of households, positive and negative changes to families can reverberate to affect the vulnerability of children. If women's land rights are not strongly defined, documented, and enforced, then food security and nutritional outcomes for children can be adversely affected.

Box 6.4 How Women's LTS Matters

Cited in USAID Fact Sheet (USAID, 2016):

- Ethiopia: land allocated to women decreased household food insecurity by 36%.
- Nicaragua and Honduras: increases in female landholdings—increases in food expenditure.
- Nepal: women own land—children are 33% less likely to be severely underweight.

Cited in Espinosa (2014):

- Ethiopia: land certification—increased caloric intake—more for female-headed households.
- Tanzania: when women have PR to land, higher incomes, and savings.
- Zambia: children whose families lost land received 11% fewer calories.

Cited by IFPRI (2019):

- Kyrgyzstan: impacts on child height.

Shocks and Hazards

Stability over time in the availability, access, and utilization of food is an important element of food security. LTS issues can interact with natural and man-made shocks and hazards (e.g., conflict, natural disasters, and environmental degradation) and decrease food security for already vulnerable people.

Freudenberger et al. (2019) provide a conceptual model to understand such dynamics and illustrate how they are playing out today in the rise of violent extremism in Africa's Sahel region. In this model, both resource governance and environmental change drive food insecurity dynamics. Weak resource governance aggravates or trigger drivers of violent extremism. Drivers of violent extremism perpetuate the impacts of weak land and resource governance, triggering additional land and resource governance challenges in a kind of negative spiral. In this context, environmental triggers exacerbated by climate change and population growth (and consequent increases in resource scarcity and migration or displacement) serve as contributing factors intensifying land governance challenges, impacts of weak land governance, and drivers of violent extremism. Importantly, the authors suggest that it will be hard to mitigate and reverse these dynamics without holistic solutions that explicitly take into account challenges to strong LTS.

Natural shocks and hazards including both discrete shocks (e.g., earthquakes, hurricanes, floods, fires) and longer-duration hazards (e.g., droughts, rainfall variability, desertification, soil degradation, wildlife loss) can increase near- and long-term food insecurity. Although each discrete shock or hazard is a stochastic event, contemporary research and discussions on shocks, hazards, and food security commonly recognize that the severity and frequency of environmental shocks are strongly linked to climate change and that LTS is critical for creating resilience to climate change-related disasters (the UN SDG Knowledge Platform; reporting on progress on Goal 2 in 2018).

For instance, fruit and vegetable production, a key component of healthy diets, is directly vulnerable to the effects of climate change by increasing uncertainty around crop production from shocks and hazards. Declines in yields and crop suitability are projected under higher emission scenarios, especially in tropical and semi-tropical regions. Heat stress reduces fruit set (i.e., the process through which flowers become fruit and the fruit size is determined) and speeds up the development of annual vegetables, resulting in yield losses, impaired produce quality, and increasing food loss and waste. While some projections indicate that longer growing seasons will enable a greater number of plantings to be cultivated and can contribute to greater annual yields, some fruits and vegetables need a period of cold accumulation to produce a viable harvest, and warmer winters may constitute a risk. Systems for LTS now must factor in these risks in land allocation, markets, and management.

Further, shocks and hazards can exacerbate food insecurity through increases in food prices. Global crop and economic models have projected a 1–29% cereal price increase by 2050 due to climate change, which would impact consumers globally through higher food prices and reduced purchasing power. Low-income consumers are particularly at risk from higher food prices (IPCC 2019).

The extent that reduced caloric intake from increased food availability leads to a heightened risk of hunger varies according to the projected pathway of socioeconomic change under climate change. However, all models projected by IPCC predict increases in the risk of hunger, with the median projected increase in the population at risk of insufficient caloric intake between 6% and 12% by 2050 compared to a no climate change reference scenario. These median percentages imply 8–80 million (full-range: 1–183 million) additional people at risk of hunger due to climate change by 2050 (IPCC 2019). The way land is allocated and used is a critical factor for determining which of these scenarios is actually realized by 2050.

While increased CO_2 is projected to be beneficial for crop productivity at lower temperature increases, it is projected to lower nutritional quality. Distributions of pests and diseases will change, negatively affecting production in many regions. Given increasing extreme events and the interconnectedness of global food and economic systems, risks of food system disruptions are growing.

The lack of effective land governance and weak LTS negatively impacts people's ability to effectively manage land for long-term sustainability, specifically for climate adaptation and mitigation. Lack of recognition and violations of property rights can increase vulnerability and decrease adaptive capacity. These issues are especially critical for customary and community forms of land tenure. According to the Rights and Resources Initiative (Rights and Resources Initiative, 2015), local communities and Indigenous Peoples hold significant areas of the earth's land under customary land systems (as much as 65%), largely undocumented or unrecognized under statutory law (only 10%). Although these communities are increasingly recognized for their long-term stewardship of the earth's natural resources (USAID, 2018), challenges for broader recognition remain.

Clarification of property rights to improve LTS can provide the security and adaptability needed for long-term resilience in food systems and supporting ecosystem services. Good land governance can strengthen the "menu" of land management options and align incentives for sustainable agriculture, including the improved management of cropland and grazing lands and sustainable forest management. In many cases, this is possible without requiring the reallocation of land rights. The efforts to create a type of "Great Green Wall" to slow or reverse land degradation in the Sahel, for example, depend importantly on the clarification of property rights for smallholders and pastoralists to land, trees, and water.

A wide range of adaptation and mitigation responses to climate change are heavily conditioned by LTS, and their success depends upon creating clear property rights to serve as the legal and economic basis for distributing subsidies, sanctioning non-compliance, and incentivizing "micro" level processes that support sustainable agriculture (dimensions 3 and 5). These include preserving and restoring natural ecosystems, such as peatland, coastal lands and forests, biodiversity conservation, reducing competition for land, fire management, soil management, and most risk management options (e.g., use of local seeds, disaster risk management, and risk-sharing instruments).

Shocks and hazards from institutional changes in landholding systems are also important. For instance, LSLBI may be a type of man-made shock with potentially significant implications for food security if the movement toward responsible investment in agriculture that respects the property rights of local smallholders and communities is not achieved. Soil degradation and investments in soil conservation are other areas in which the institutional arrangements for LTS may be crucial for avoiding adverse consequences for food security. In Malawi, for example, soil degradation associated with climate change is a significant problem, and soil conservation practices have been shown to be weakened by land tenure arrangements which tend to lead to short-term rental contracts and gender-biased inheritance practices (Lovo, 2016).

Illustrating the Case: Food Security Challenges and Patterns of Land Tenure in South Asia

According to the World Hunger Report 2017, South Asia ranks as the region facing the most severe food security challenges. For instance, India slipped to 100th place for food security after being ranked 55th in 2014. The interconnected challenges of increasing sustainable agriculture, mitigating and increasing resilience to climate change, and increasing food security are particularly pronounced in South Asia (Bandara & Cai, 2014). About 600 million South Asians live under the World Bank poverty line of less than US$1.25 a day, the majority of whom depend directly or indirectly on agriculture (Hertel et al., 2010).

Transitioning to agricultural development that addresses climate change mitigation and resilience and maintains food security is clearly a priority across South Asia. However, under the Paris Accord, the Nationally Determined Contributions of South Asian countries, vis-à-vis the agriculture sector, focus is more on technology and market-based solutions rather than addressing inherent structural weaknesses related to LTS (Amjath-Babu et al., 2019). In our view, understanding patterns within the PLR helps provide a logical answer to why hunger and malnutrition persist in South Asia, particularly in India. Addressing food security concerns and tackling the looming threat of climate change make land tenure reforms necessary (Padhee & Joshi, 2019).

More than half of the farmland in the so-called "poverty square" of South Asia consists of marginal and small farms of less than one hectare. Most of these farmers are sharecroppers and tenants whose rights are unrecorded—informal or concealed (when they are not legally allowed) because of ineffective and restrictive land leasing legislations (Appu, 1975). Small and marginal farmers increasingly lease-in land (i.e., acquire access to land by leasing it for use from another, usually larger landholder) in low-productivity regions, whereas in agriculturally advanced areas the trend is toward reverse tenancy (i.e., the poorer landowner rents out land to richer tenants) (Kumar et al., 2017; Patel & Mishra, 2019).[1] In this case, low-productivity regions refer to areas where the agriculture practiced is rain-fed, subsistence, and mostly food-crop oriented. Advanced areas refer to those with irrigation and a greater orientation to commercial crop production.

Leasing-in land can support a household's immediate food availability, and leasing-out can generate income to support food access. In some places, short-term tenancy forms cover as much as 25% of the gross cultivated area, although the true number is unknown as these arrangements

[1] Area leased-in is defined as land taken on lease from others without any permanent rights of possession for the lessee. Land may be leased-in for fixed produce, fixed money, share of produce, or usufructuary mortgage.

Government of India (1986) Agriculture Census 1985–1986. http://agcensus.nic.in/document/ac8586/reports/Annex_Concept%20&%20Defination_8586.pdf

are often underreported (Planning Commission 2013). However, land access can be precarious, and short-term tenancies have been correlated with decreasing land productivity and degradation, adversely affecting food security over time. Research on these issues in India highlights that insecure rights and the risk of not capturing the full value of durable investments make farmers less likely to conserve land or make productivity-enhancing investments (FAO, 1994, Deininger et al. 2013, IFAD, 2011, Osbahr et al., 2010).

In the absence of legal recognition of tenancy, tenants lack access to a range of public service entitlements, such as formal credit, insurance, crop-loss compensation, minimum support price, fertilizer subsidy, loan waivers, and direct cash transfers (Raju, 2019). This can increase uncertainty and vulnerability to food insecurity and can affect a household's food production and their ability to purchase food. The fear of agricultural lands falling into the hands of the sharecroppers after a specific period due to land reform legislation also contributes to many absentee landowners keeping lands fallow, affecting overall food production (Ranganathan & Pandey, 2018).

India's national policy think tank, the National Institute for Transforming India (NITI Aayog), has recognized land leasing as an "economic necessity" in the Indian context, in contrast to earlier reformist legislation which banned it. In 2016, NITI Aayog prepared model legislation to formalize and improve tenancy arrangements to encourage policy reform in Indian states. Legalization of leasing will encourage more land to be leased and will allow leases to be formally documented, expanding access to a range of public service entitlements for informal tenants.

Despite efforts to increase LTS for informal tenants, new data illustrate how LTS constraints for food security continue to impact rural households (Strengthening Adaptive Farming in Bangladesh, India, and Nepal [SAFBIN]). The data come from a 2018–2019 survey of 1145 households of small-scale producers from 95 villages in 12 districts in India, Nepal, Bangladesh, and Pakistan. This program to strengthen adaptive farming is operational in these 95 villages, covering approximately 5000 smallholder farming households.

In the sample, one-third of households do not own land. Rather, they rely on leasing or public land possession for farming. Leasing-in land is more prevalent among farmers with the smallest farm sizes (<0.4 ha), of which half rely on leased-in land. Half of the adult members of farm families lack landownership, regardless of records or informal means of access (by practice or custom). Two-thirds of this half are women. Women farmers are also highly marginalized across all forms of tenure.

While almost all those who own land have formal land records, informality is common for the majority of leased tenure and public land possessions. More than two-thirds of the farmers perceive they have secure land tenure with formal records, while only two-fifths of those without records report perceptions of secure tenure. Most surveyed farmers reported having secure tenure despite tenure diversity and complexities related to different forms of tenure and informal leasing arrangements. Those having leased-in land and the smallest parcels report lower LTS.

Most of the surveyed households access land under more than one form of tenure. More than four-fifths (86%) reported cultivation on their own land, and 21% of these farmers additionally lease-in land from others. Very few farmers (3%) leased-out land, and only about 7% cultivate on public land (*de jure* government land). With the legality of tenancy remaining confounded and public land cultivation considered encroachment, LTS for about one-third of smallholder farmers remains insecure. There are differences in the distribution of these forms of tenure, however, across countries in the region (see wide ranges in parentheses within Table 6.2). In Bangladesh, more than half of the smallholder farmers

Table 6.2 Forms of tenure among farm households across land size class categories for Bangladesh, India, Nepal and Pakistan

Form of tenure/farm size	<0.4 ha	0.4–1 ha	1–2 ha	>2 ha	All
Own land	46%	29%	17%	5%	68%
	(8–94)	(6–43)	(0–42)	(0–21)	(58–81)
Leased-in	55%	26%	14%	5%	24%
	(4–93)	(7–44)	(0–50)	(0–22)	(14–29)
Leased-out	56%	32%	9%	3%	3%
	(0–100)	(0–50)	(0–100)	(0–4)	(0–6)
Possession on government land	87%	4%	7%	1%	6%
	(0–100)	(0–17)	(0–42)	(0–100)	(0–12)

reported leasing or using public land (41% lease-in and 18% use public land, respectively), whereas in India the figures are 18% (14% lease-in and 4% public land, respectively). Gender-based differences are also prevalent: women reported control of only 11% of the land compared to men who reported controlling 89% of the land.

In looking at the SAFBIN data related to the SDG land indicators for the four countries, which focus on the individual rather than the household, every second adult member of smallholder farming households does not legally or customarily own any land. Among the countries, ownership rights are lowest in Pakistan (31% own land) and highest in Nepal (78% own land). Here again, the data indicate gendered differences in landownership, with two-thirds of women lacking landownership compared to one-third among men. Gender differences are highest in Pakistan.

As shown in Table 6.3, one in four adult members of smallholder farming households has documentary evidence for their landownership, while another one in five own land through customary means. Documentary evidence is more frequent among farmers in India (23% have formal documentation compared to 7% with customary ownership) and Pakistan (24% and 9%), while in Nepal more farmers (34%) own land through customary means compared to 26% with documented ownership.

Among the smallholder farm households surveyed by SAFBIN, more than half (58%) have government-authorized legal documents. About one-third also use land without formal contracts. However, almost all (96%) of the farmers who own land have proper records (Table 6.4). On the contrary, more than 80% each of leased-in and leased out holdings

Table 6.3 Basis of landownership of adult individuals in farming households in Bangladesh, India, Nepal and Pakistan

Landownership basis	Average %	Range	Adult male (%)	Adult female (%)
As per formal documentary evidence/land record	25	20–32	19	6
Local customs or practices or will	18	7–34	13	5
Not a landowner	50	32–69	16	34
No response	7	1–16	5	2

Table 6.4 Rate of formality by type of land access in the SAFBIN survey

	Government-authorized documents including notarized contracts (%)	Oral contracts and white paper contracts (%)
Own land	96	4
Leased-in *(note: 5% no responses)*	11	84
Leased-out	17	83
Possession on government land	18	82

are through oral contracts. Informality (i.e., oral, white paper, or no contracts) is common practice for about two-fifths of the smallholder farmers owning less than one hectare.

Despite the diverse, informal, and unrecorded tenure types, 81% of smallholder farmers feel their holdings are secure, while 12% feel a threat of loss in the next five years. However, disaggregation across tenure types reveals that the threat of losing land is higher for leased-in farmers (53%) and for those possessing public land (20%). Similarly, for farmers holding less than 0.4 ha, even more feel insecure (74%). Limited access to land and precarious tenure arrangements (undocumented leasing) are generally associated with a greater risk of food insecurity.

Connecting the Dots between the LTS and Food Security in South Asia

Using the information on dietary diversity, farm composition (i.e., field crops, trees, livestock, compost unit, and crops cultivated), the SAFBIN data indicate linkages between LTS, food security, and farm sustainability patterns. At the household level, survey data indicate that under all tenure forms households have similar *food access*. Households reported having access to cereals (almost 100%), pulses (~90%), and vegetables (~97%). However, *land access* is a constraint for pulses and vegetables when looking only at leasing-in farmers compared to farmers that own land. Further, there is variation in food access based on the household member.

Individual household members with land documentation had a higher frequency of food access across all food groups compared to household members without formal documentation. Dietary diversity for farmers that only leased-in land or occupied government land is poorer than for farmers whose holdings are owned or leased-out. Leased tenures and lack of land documents make farmers and their family members more vulnerable to nutritious food access, suggesting that LTS is one of the structural keys for addressing hunger and malnutrition in South Asia.

Regarding sustainable land usage, crop diversity—here a proxy for sustainable agricultural production—is lowest for farming under the possession of public land, increasing gradually from farmers with leased-in land (only), owned land, leased-out land, and other combinations of landownership. Diversity of land uses (e.g., field crops, trees, livestock, and fisheries) is highest among farmers with the smallest landholdings, gradually decreasing as landholding size increases. Farm diversity is found to increase as tenure type moves from leased land, public land possession, landownership, to leased-out land.

Informal leasing also seems to contribute to increased fallowing of cultivable lands in many states in India, which affects food production. Fragmented and smaller landholding size is often cited as a reason for reductions in farmers' ability to adapt to climate change (Aryal, 2019). Yet, smaller farms with ownership rights in South Asia seem to have a penchant for crop diversification as a strategy for resilience and are more likely to apply it when they have tenure security, consistent with the SAFBIN data. Apart from affecting food security, lack of attention to improving LTS therefore also hinders upscaling of locally known adaptation measures.

The insecurity demonstrated by farmers with smaller parcels and leased tenure forms across the four countries seems to highlight why there is increasing demand to formalize leasing in India. The extent to which those with ownership rights lack up-to-date and accurate land records is also notable. Both situations limit the ability of farmers to access credit and entitlements for optimal food and crop production, even if farmers perceived their tenure to be secure. Prioritizing a land reform agenda, particularly land leasing legislation and updating land records, is likely an

Box 6.5 Improving LTS—Including for Women—Offers Hope in West Bengal

A land-allocation and registration program that focused on women had positive impacts on outcomes expected to lead to future food security: beneficiary households reported stronger security and were more likely to take loans for agricultural purposes, to invest in agricultural improvements, and to involve women when making decisions related to food and . The same types of variation as seen in SAFBIN by farm size were observed and the effects were larger if women's names were on a recorded on the land title (Santos et al., 2013).

important path to increasing incomes and improving food security for smallholders, at least in India (Padhee & Joshi, 2019), if not across South Asia. As with the general case for connecting LTS and food security, gender matters (see Box 6.5).

From Global to Local: Evidence Supports Linking Efforts on LTS and Food Security

The discourse on food security within the international development community is abundantly clear on its assertion that LTS is integrally linked to achieving food security. We believe that there is support for this assertion from a dispersed but compelling body of global evidence. Stated commitments by international bodies to improve LTS and achieve food security for all are increasing. Attention to the interconnectedness of these two global challenges also appears to be increasing. While gaps in, and challenges for, research remain, we believe the case for action to strengthen LTS for food security is already sufficient. LTS, sustainable agriculture, and food security are all multidimensional and vary across contexts. The research into the inter-relationships among them is both siloed by theme and dispersed across disciplines with few truly multidisciplinary approaches. More study is certainly needed to strengthen the evidence for each of the assumptions in the causal chain within LTS and its connections to food security across diverse geographies and institutional contexts. But we believe the evidence assembled here in a holistic and accessible synthesis provides compelling support for a call to immediate, integrated, and widespread action to strengthen LTS in support of food security, even while more research is underway.

References

Amjath-Babu, T. S., Aggarwal, P., & Vermeulen, S. (2019). Climate action for food security in South Asia? Analyzing the role of agriculture in nationally determined contributions to the Paris agreement. *Journal of Climate Policy, 1*, 283–298.

Appu, P. (1975). Tenancy reform in India. *Economic and Political Weekly, 10*(33/35), 1339–1375.

Aryal, J. S. (2019). Climate change and agriculture in South Asia: adaptation options in smallholder production systems. *Environment, Development and Sustainability*, 1–31.

Baltissen, G., & Betsema, G. (2016). *Linking land governance and food security in Africa: Outcomes from Uganda, Ghana and Ethiopia*. LANDac.

Bandara, J., & Cai, Y. (2014). The impact of climate change on food crop productivity, food prices and food security in South Asia. *Economic Analysis and Policy, 44*(4), 451–465.

Bowen, D., & Ngeleza, G. (2019). *MCC land sector cost-benefit analysis guidance*. The Millennium Challenge Corporation.

Cipollina, M., Cuffaro, N., & D'Agostino, G. (2018). Land inequality and economic growth: A meta-analysis. *Sustainability, 10*(12), 46–55.

Deininger, K. (2003). *Land policies for growth and poverty reduction*. The World Bank.

Deininger, K., & Byerlee, D. (2011). *Rising global interest in farmland: Can it yield sustainable and equitable benefits?* The World Bank.

Deininger, K., & Squire, L. (1998). New ways of looking at old issues: Inequality and growth. *Journal of Development Economics, 57*(2), 259–287.

Deininger, K., Jin, S., & Nagarajan, H. (2009). Land reforms, poverty reduction, and economic growth: Evidence from India. *The Journal of Development Studies, 45*(4), 496–521.

Deininger, K., Songqing, J., & Yadav, V. (2013). Does sharecropping affect long-term investment: Evidence from West Bengal's tenancy reforms. *American Journal of Agricultural Economics, 95*(3), 772–790.

Espinosa, D. (2014). *Secure Land Rights: Food Security Depends On It.* Focus on Land In Africa.

FAO. (1994). *Land degradation in South Asia: Its severity, causes and effects upon the people.* FAO.

FAO; IFAD; UNICEF; WFP and WHO. (2019). *The state of food security and nutrition in the world.* FAO.

Fort, R. (2007). Land inequality and economic growth: A dynamic panel data approach. *Agricultural Economics, 37*(2–3), 159–165.

Freudenberger, M., Sanjak, J., Tagliarino, N., & Thomson, N. (2019). *Climate change, land and resource governance and violent extremism: Spotlight on the African Sahel.* Tetra Tech.

Hertel, T., Burke, M., & Lobell, D. (2010). The poverty implications of climate-induced crop yield changes by 2030. *Global Environmental Change, 20th Anniversary Special Issue, 20*(4), 577–585.

Holden, S., & Ghebru, H. (2016). Land tenure reforms, tenure security and food security in poor agrarian economies: Causal linkages and research gaps. *Global Food Security, 10*, 21–28.

Hufe, P., & Heuermann, D. (2017). The local impacts of large-scale land acquisitions: A review of case study evidence from sub-Saharan Africa. *Journal of Contemporary African Studies, 35*(2), 168–189.

International Food Policy Research Institute (IFPRI). (2019). *Global food security report: Synopsis.* IFPRI.

International Fund for Agricultural Development (IFAD). (2011). *IFAD rural poverty report.* IFAD.

Jorgensen, S. L., & Siegel, P. (2019). *Social Protection in an Era of Increasing Uncertainty and Disruption: Social Risk Management 2.0; Social Protection and Jobs Discussion Paper; no. 1930.* Washington, DC: The World Bank.

Keswell, M., & Carter, M. (2014). Poverty and land redistribution. *Journal of Development Economics, 110*, 250–261.

Kumar, S., Meena, P., & Panigrahy, S. (2017). Changing dynamics of tenant farming in India—An insight from Gujarat. *Indian Journal of Economics and Development, 13*(1), 65–70.

Lawson, M., Schnepf, R., & Cook, N. (2013). *The Obama Administration's feed the future initiative.* Congressional Research Service.

Lisher-Witriol, J. (2019). *Guidelines for impact evaluation of land governance interventions.* Global Land Tools Network and the International Fund for Agricultural Development.

Lovo, S. (2016). Tenure insecurity and Investment in Soil Conservation. Evidence from Malawi. *World Development, 78*, 219–229.

Maxwell, D., & Wiebe, K. (1998). *Land tenure and food security: A review of concepts, evidence and methods, research paper no. 129*. The Land Tenure Center, University of Wisconsin.

Maxwell, D., & Wiebe, K. (1999). Land tenure and food security: Exploring dynamic linkages. *Development and Change V., 30*(4), 825–849.

Muraoka, R. J., & Jayne, T. (2018). Land access, land rental and food security in Kenya. *Land Use Policy, 70*, 611–622.

Nkomoki, W. B. (2018). Adoption of sustainable agriculture practices and food security threats: Effects of land tenure in Zambia. *Land Use Policy, 78*, 532–538.

Osbahr, H., Twyman, C., Adger, W., & Thomas, D. (2010). Evaluating successful livelihood adaptation to climate variability and change in southern Africa. *Ecology and Society, 15*(2), 1.

P. R. Shukla, J. S.-D.-O. (2019). *Climate change and land: An IPCC special report on climate change, desertification, land degradation, sustainable land management, food security, and greenhouse gas fluxes in terrestrial ecosystems*. IPCC (In Press).

Padhee, A., & Joshi, P. (2019). *Why India needs a land leasing framework*. IFPRI Blog Post.

Patel, R., & Mishra, D. (2019). Agrarian transformation and changing labour relations in Kalahandi, Odisha. *Journal of South Asian Development, 14*(3), 314–337.

Paudel, S., Deng, W. P., Khatiwada, J., Zhang, J., & Su, Y. (2017). Household Livelihood Strategies and Implication for Poverty Reduction in Rural Areas of Central Nepal. *Sustainability.*

Raju, Y. (2019, January 24). Tenant farmers being left high and dry. *The Hindu Business Line*, pp. https://www.thehindubusinessline.com/opinion/tenant-farmers-being-left-high-and-dry/article26081913.ece

Ranganathan, T., & Pandey, G. (2018). Who leaves farmland fallow and why? An empirical investigation using nationally representative survey data from India. *The European Journal of Development Research, 30*(5), 914–933.

Santos, F., Fletchner, D., Savath, V., & Perterson, A. (2013). *Can Government-Allocated Land Contribute to Food Security? Intrahousehold Analysis of West Bengal's Microplot Allocation Program.* Washington, DC: IFPRI Discussion Paper 01310.

Sherraden, M. (1991). *Assets and the poor: A new American welfare policy*. M.E. Sharpe.
The Rights and Resources Initiative. (2015). *Who Owns the World's Land?* Washington, D.C.
USAID. (2016). *Fact sheet: Land tenure and food security*. USAID.
USAID. (2018). *Draft USAID policy on indigenous peoples issues*. USAID.

7

A Complex Relationship: Large-Scale Land Acquisitions and Land Tenure Security

A Global Overview and Insights from Zambia

Kerstin Nolte, Chewe Nkonde, Paul Samboko, Raoul Herrmann, Sarah Holzapfel, Henry Machina, and Mitelo Subakanya

Land Tenure Security and Large-Scale Land Acquisitions

Over the last decade, we observe an emerging interest in agricultural land globally. Due to the opaque nature of such large-scale land acquisitions (LSLAs), it is difficult to assess the true extent of the phenomenon, let alone speak of the impacts. However, data from the Land Matrix Initiative

K. Nolte (✉)
Institute of Economic and Cultural Geography, Leibniz University Hannover, Hanover, Germany
e-mail: nolte@wigeo.uni-hannover.de

C. Nkonde
Department of Agricultural Economics and Extension, University of Zambia, Lusaka, Zambia

M. B. Holland et al. (eds.), *Land Tenure Security and Sustainable Development*,
https://doi.org/10.1007/978-3-030-81881-4_7

estimates that more than 1500 LSLAs on more than 42 million hectares in low- and middle-income countries have been taken by transnational investors alone. There is also growing evidence that domestic investors play a key role in land acquisitions (Lavers, 2012; Nolte & Sipangule, 2017), although information on the extent of their role remains thin. This emerging evidence reflects the rising interest in LSLAs by international and domestic investors to acquire land in low- and middle-income countries for agricultural (and other) purposes.[1] Global factors that have contributed to the increase in LSLAs include the food price spike in 2008, a rising global population, changing consumption patterns, and an increased demand for agrofuels (World Bank, 2010). At the national level, a conducive investment climate and the development of land markets play an important role (Sambo et al., 2015; Samboko et al., 2018).

Due to the persisting intransparency surrounding such acquisitions, the existing information on LSLAs by transnational and domestic investors from the Land Matrix is likely just the tip of the iceberg of a much broader trend in land acquisitions. Growing evidence shows that LSLAs happen all over the world, with hotspots in Southeast Asia, Eastern Europe, Brazil, and sub-Saharan Africa—with sub-Saharan Africa being the most targeted region and the focus of this chapter. Investors from industrialized countries play a disproportionate role, including major investor countries such as the United States and the United Kingdom. The most common investors are from Western Europe, followed by

[1] We will focus on agricultural purposes in this chapter, bearing in mind that land is also acquired for other reasons, including (but not limited to) mining, tourism, and conservation.

P. Samboko
Disaster Management and Mitigation Unit, Office of the Vice President, Government of Zambia, Lusaka, Zambia

R. Herrmann • S. Holzapfel
German Development Institute/Deutsches Institut für Entwicklungspolitik (DIE), Bonn, Germany

H. Machina • M. Subakanya
Indaba Agricultural Policy Research Institute, Lusaka, Zambia

Southeast Asia. Global South investors show a preference for investing in their own region (Nolte et al., 2016).

Arguments for and against the growing trend in LSLAs can be polarizing, as questions remain about whether LSLAs can be implemented in an economically, socially, and environmentally sustainable way. Proponents have argued for the renewed interest in the long-neglected agricultural sector and praised the associated development potential of LSLAs, while opponents refer to the phenomenon as "land grabbing," hinting at rather weak protection of land rights and adverse outcomes for affected populations. Regardless, LSLAs are directly linked to one of the key challenges of our times: *sustainable* development. Understanding the complexities of how LSLAs occur is a key step in understanding its impacts on the well-being of local populations and the use of natural resources, and how policies may facilitate an equitable and fair process.

LSLAs take place in complex land governance systems, and these systems can shape the way land acquisitions are implemented and affect the land tenure security of local populations. This is because whether rights to access and use land are being upheld when investors enter depends on the land governance system. Decreases in land tenure security itself can ultimately result in adverse outcomes for local populations. For instance, in many target regions—especially in sub-Saharan Africa where customary tenure regimes coexist with statutory tenure (Alden Wily, 2011; Boone, 2014)—documentation of land rights is poor, and overlapping land rights can lead to conflicts (Lund, 2008). If LSLAs happen in such contexts, land tenure security for local land users is likely to be particularly weak. Detailed case studies across sub-Saharan Africa document how investors negotiate LSLAs in the specific institutional context (Bottazzi et al., 2016; German et al., 2013; Nolte & Väth, 2015). These studies show how some investors use institutional weaknesses for the advantage of implementing their projects.

This chapter provides a global overview on the relationship between land tenure security and LSLAs and highlights the Zambian context to provide an illustrative example of how land acquisitions affect sustainable development. We focus on three topics: first, the role of land tenure security in attracting investors; second, the role of land tenure security in mitigating adverse impacts of LSLAs; and third, the impacts of acquisitions on land tenure security. We close with discussing what role land tenure security can play in implementing LSLAs in a more sustainable fashion.

Global Overview

Land tenure security plays an important role in LSLAs in three ways. First, it affects the locational choice of investors. Despite the fact that investments typically require stable institutions, the LSLA literature discusses whether weak land tenure security attracts investors (Arezki et al., 2013; Lay & Nolte, 2018). We start with a quick visual impression and plot the number of LSLAs against the Global Index of the Governance Context for Land Tenure Security (GC-LTS) as a (national) measure of land tenure security for low- and middle-income countries.[2] The GC-LTS represents contextual factors of the governance context for land tenure security and is scaled from 0 to 1 with higher values signaling stronger land tenure security. The size of the circles is determined by the size of all LSLAs in a given country. Figure 7.1 does not show a clear relationship: land acquisitions occur in countries with different GC-LTS values. Most land acquisitions occur in countries with medium GC-LTS values (between 0.4 and 0.6). Acquisitions in countries with lower GC-LTS are fewer (and smaller), while very few acquisitions take place in countries with higher GC-LTS.

Empirical findings on the role of land tenure security in the locational choice of land acquisitions are rather scant. The literature typically finds that, generally speaking, institutions are an important determinant of land acquisitions, but the direction of the relationship is less clear. While Arezki et al. (2013) find that weak land tenure security is associated with more land acquisition projects, Lay and Nolte (2018) do not find such a straight-forward relationship. In contrast to the literature on foreign direct investment (FDI), the authors do not find a positive relationship with institutions and conclude that some deficiencies in specific institutions, such as corruption control, are tolerated by investors. Raimondi

[2] The GC-LTS provides information at the national level; hence variations within countries are not reflected in it.

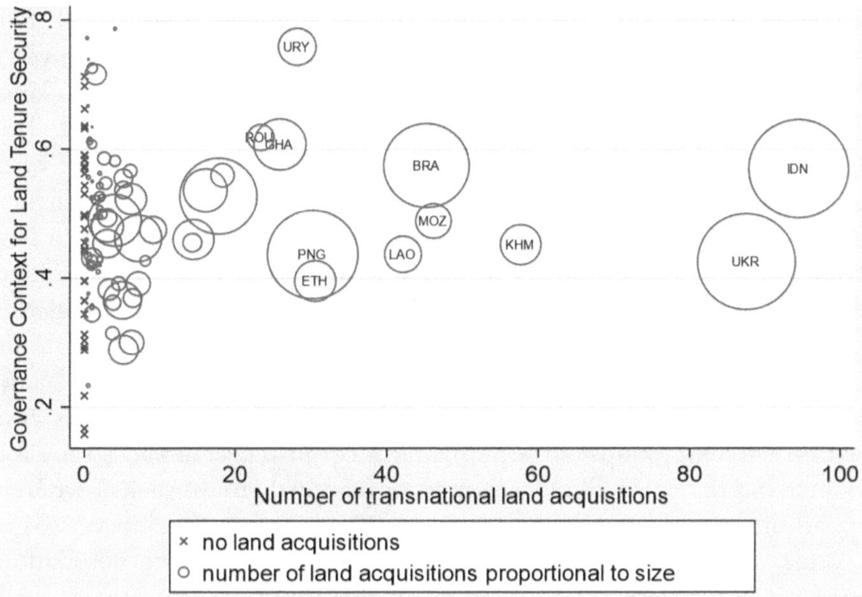

Fig. 7.1 Relationship between land tenure security and land acquisitions. (Source: data from Land Matrix Initiative (data as of February 8, 2018) and the Governance Context for Land Tenure Security (Kelly et al., 2017; Land Matrix, 2019). Each observation corresponds to a low- and middle-income country (the same set of countries that are considered in the Land Matrix data; countries with more than 20 acquisitions have a label (ISO 3 code))

and Scoppola (2018) take a different stance and find that the institutional distance (i.e., different institutional structures) between the target country and the origin of the investor matters. Countries with weak governance contexts for land tenure security would then invest in countries with similar contexts. However, patterns differ across geographical regions, and Africa follows a pattern of its own. This latter finding is likely to explain the ambiguous insights from Fig. 7.1: the relationship between land acquisitions and land tenure security is complex, certainly not linear, and deeply rooted in local institutional practices.

Second, land tenure security can function as a mitigation channel for impacts of LSLAs: whether land acquisitions occur in a context of weak or strong land tenure security makes a difference for the effects of LSLAs on local communities and the environment (German et al., 2018). For instance, Herrmann (2017) argues for the case of Tanzania that investments implemented on former village land may be more prone to land conflicts compared with investments on former government land. The diversity of outcomes of LSLAs are widely discussed in the literature with a focus on socio-economic outcomes (Ali et al., 2016, 2019; Baumgartner et al., 2015; Deininger & Xia, 2016; Herrmann, 2017), and only a few studies on the environmental outcomes (Johansson et al., 2016; Shete et al., 2016; Zaehringer et al., 2018). While a comprehensive and full understanding of the impacts of LSLAs is still lacking, evidence so far suggests overall impacts are rather negative (Lay et al., 2021a). Many scholars argue that the impacts tend to be heterogeneous across and within different groups of the population. For instance, impacts differ for poorer and richer smallholder farmers or marginal groups such as women or ethnic minorities (Behrman et al., 2012; Borras & Franco, 2013; Cotula, 2013). Consequently, while certain groups of the population might well benefit from an investment, this is not true for others—and marginalized groups might even lose out. A key aspect in determining the outcomes of an LSLA is the "inclusiveness" of the business to be established (e.g., a commercial farm), which in turn is shaped by the local land governance system (German et al., 2018). The idea is that businesses have greater incentives to work with rural people who have secure land tenure and include them in their commercial ventures. This, in turn, is thought to improve the well-being of rural people.

Third, LSLAs can impact land tenure security. For instance, land acquisitions may displace local land users and/or deprive them of their access to land. Several case studies of land acquisitions from across the world report on the loss of access to land and displacements. Bottazzi et al. (2018) report a case in Sierra Leone where land accessible to households for agriculture was reduced by 50 percent. In Cambodia, Neef and Singer (2015) highlight that LSLAs are responsible for a large number of displacements. Although quantitative evidence on displacements is scarce, Land Matrix insights suggest that if displacements occur, many people are affected. Given that most of the land acquired is not idle land, the loss of access to land and displacements that occur are likely frequent (Nolte et al., 2016). Moreover, due to the presence of new investors, the community's perception of land tenure security can change. The arrival of large-scale farms is often accompanied by a heightened sense of uncertainty, land scarcity, or tenure insecurity for smallholders (Cotula, 2011). Finally, the arrival of new investors may bring hidden deficiencies in the land governance systems and land conflicts to the surface, thus increasing calls for reforms that put further pressure on land governance systems (Bottazzi et al., 2016). Many land governance systems are unprepared for the relatively sudden appearance of wealthy investors. This may, in turn, catalyze calls to reform the land governance system, which can affect the land tenure security of local communities (Nolte & Väth, 2015).

In the following section, we will shed light on Zambia to further discuss the relationship between land acquisition and land tenure security.

Large-Scale Land Acquisitions and Tenure Security Contextualized: A Case of Zambia

Changes in the legal framework of Zambia's land governance system have been a major supporting factor of LSLAs. Although Zambia has a long history of large-scale farms that have coexisted alongside smallholder communities, the rise in demand for land since 1995 is unprecedented (Chu, 2013; Chu, Young, & Phiri, 2015; Nolte, 2014). Zambia has a land tenure system marked by a strong customary system, which in 1964 comprised 94 percent of the land area, and a statutory system that comprised the remaining 6 percent of state land. According to estimates by Sitko and Chamberlin (2016), customary land is down to about 54 percent of the total land area today. A major change came with the 1995 Lands Act, which, among other goals, strengthened property rights of titleholders on state land and eased land ownership by foreigners. Since its enactment, land acquisitions have increased rapidly in Zambia (Nolte, 2014). Besides the ability to acquire state land directly from the government for investment purposes, foreigners were also able to acquire large tracts of land directly from traditional leaders and convert it to leasehold tenure, typically granted for 99 years and renewable thereafter.

The recent surge in the acquisition of land in Zambia can be differentiated as follows. The first type comprises foreign-owned firms involved in agriculture, manufacturing, and extractive industries characteristically covering an area of 200 hectares or larger per acquisition (Land Matrix, 2019). About 26 deals with foreign-owned firms were concluded during 2000–2015, with the total size under contract estimated at approximately 390,000 hectares, which is more than the total area of Rhode Island (Land Matrix, 2019). The pace of these deals increased after 2011, supporting the argument that the recent LSLA phenomenon was in part driven by the 2008 global food crisis. The second type, the rise of domestic investor farms (those not exceeding 100 hectares landholding each—or medium-scale acquisitions), encompasses land acquisitions that have been under the radar in the development discourse but are shifting land distribution in Zambia from a more egalitarian broad-based farm structure to one that is highly concentrated (Jayne et al., 2014, 2016). These investments are partly a consequence of the increase in demand for land by the country's middle-class who see agriculture as a viable investment opportunity. Domestic investor farms are increasing at an extraordinary pace. Estimates suggest that the total area under the control of these farms exceeds that of foreign and previously established domestic large-scale holdings combined. Further, domestic investor farms now control more land than small-scale farms who represent the majority of farms in Zambia (Jayne et al., 2014).

Despite the interest that the emerging medium-scale farm sector has recently generated, we focus our discussion on land tenure security associated with LSLAs. However, as often both forms of increased demand for land happen in the same regions, it is likely that synergies between both forms exist (Lay et al., 2021b).

Does Weak Land Tenure Security Attract Investors in the Zambian Context?

Zambian state land can be leased for 99 years (with the option to renew) and is supported by documentation in the form of a certificate of title. Generally, ownership of land with certificate of title is perceived as a more secure form of land tenure. Customary land, on the other hand, is held in trust by traditional leaders, and most smallholder communities are domiciled in areas that are under this tenure system.

In the Zambian context, the increase in LSLAs was only possible after the change in the land governance system with the Lands Act of 1995, which made it easier for investors to acquire land. Weaknesses in the system and a lack of enforcement have been identified, including a strong role of powerful individuals in the acquisition process, a lack of consultation with communities, weaknesses in the process of Environmental Impact Assessments, and the lack of a dispute resolution mechanism (Chu, Young, & Phiri, 2015; Henley, 2017; Nolte, 2014). These weaknesses impose few restrictions on investors—especially for land acquired under customary land tenure—and reduce transaction costs for investors. This could in turn attract more investors with low accountability.

Empirical evidence on this relationship for the case of Zambia is lacking. Our own interviews with investors in 2010 and 2011 show that investors rather complain about weak institutions and long processes for environmental impact assessments. No investor mentioned weak land tenure security as a determinant for land acquisitions. However, investors willing to talk to researchers are likely to be a biased sample of investors, as investors who might be attracted by weak tenure security may be less likely to speak to researchers.

Diversity of Outcomes of Land Acquisitions

We still lack a comprehensive understanding of the livelihood outcomes of LSLAs, and the same is true for the case of Zambia. While experts agree that land tenure security is particularly weak on customary land due to a lack of legal titles and formal documentations (Honig & Mulenga, 2015; Nhandu, 2017), it is less clear how different land tenure arrangements affect the diversity in outcomes following LSLAs.

While the Zambian government has created a conducive environment for LSLAs in order to increase investment opportunities, critics have argued that the sources of livelihood for local communities affected by such investments are bound to be threatened (Chu & Phiri, 2015). The majority of studies indeed find rather negative impacts from LSLAs, such as the loss of access to land, increasing land scarcity, and adverse environmental effects (Milimo et al., 2011; Mujenja & Wonani, 2012; Nolte, 2014). However, studies have also found positive aspects, including job creation and an increase in farm wage income (Ahlerup & Tengstam, 2015; Mujenja & Wonani, 2012; Schüpbach, 2014), improved access to infrastructure (Milimo et al., 2011), and improved social capital for communities close to large-scale farms (e.g., people living closer to large-scale farms or working for them show more cooperative behavior) (Khadjavi, Sipangule, & Thiele, 2019).

A key finding from these studies is that impacts are diverse—that is, different parts of society are affected differently by land acquisitions. The diversity of impacts can mainly be explained through different employment opportunities and opportunities from contract farming schemes, which are agreements between farmers and investors over production and sale of agricultural produce that provide farmers with market access, access to credit, technical advisory services, and inputs. For instance, impacts differ along gender lines, with women being less likely to benefit from LSLAs as they are mainly employed in lowly paid seasonal jobs (Matenga & Hichaambwa, 2017; Mujenja & Wonani, 2012). Impacts also differ along generational differences and wealth and poverty lines, as the youngest and poorest parts of society are often given the least paying jobs (Matenga & Hichaambwa, 2017; Mujenja & Wonani, 2012). Lay et al. (2021) observe positive spillovers in the form of increased productivity, yet farmers with slightly larger farm sizes benefit more than the smallest smallholders. Moreover, if contract farming is being implemented, it is mainly among farmers with more social, financial, and political capital who take part in those schemes, as participation is often contingent on ownership of land, giving them the ability to fully benefit from these programs (Matenga & Hichaambwa, 2017). In addition, impacts across different groups of contract farmers also differ: while dividends are good for some who are involved in the contract farming program, the distribution of these gains is uneven, both between and within households. Dividends are mainly captured by men and by external elites that have gained access to the land (Matenga, 2017).

Impacts of Large-Scale Land Acquisitions on Land Tenure Security in Zambia

In this section we focus on the impacts of LSLAs on land tenure security through the three channels: displacements, perception on land tenure security, and pressure on the land governance system. Box 7.1 portrays one specific large-scale farm in Zambia's Central Province and uses this example to discuss displacement impacts and show how the perception of land tenure security for surrounding smallholders changes.

Displacements

Studies have shown evidence of displacements due to the acquisition of land by investors (Chu, Young, & Phiri, 2015). Consultations with the communities affected by these land-based investments are limited or non-existent. While the 1995 Lands Act provides for consultation among individuals to be displaced, there is no clarity on the process of consultation (i.e., who and how the affected communities are to be consulted) (Tagliarino, 2014).

Besides the lack of consultation, there is no clear division of duties for resettlement among government bodies that take part in the resettlement process. The legal framework is unclear on who is responsible for monitoring and ensuring compliance with investor commitments (Chu, Young, & Phiri, 2015). There are also significant delays in the environmental impact assessment (EIA) process and resettlement action plan (RAP) development. Paradoxically, the EIA usually comes after the RAP, and, as such, potential adverse environmental impacts are seldom mitigated because recommendations from EIAs are rarely, if ever, incorporated into RAPs (Henley, 2017). In addition, where households are displaced, compensation is mostly inadequate and EIAs are weakly enforced, leaving everything to the investor's discretion. There are no legal provisions for compensating economic and socio-cultural losses associated with livelihood source losses, such as forests, and only loss of crops and fruit trees is considered. In general, the lack of monitoring and enforcement that the promised benefits are delivered, and the potential adverse impacts are mitigated, reinforces the power imbalance between investors and local communities.

Examples of displacements due to LSLAs in Zambia are plentiful. The creation of an oil palm project in Mpika district (north of Zambia) saw two villages losing their land for agricultural purposes, with a total of 45 households living in the land allocated to the new project being displaced and resettled (German et al., 2011). In the northwestern province of the country, Chu and Phiri (2015) report on a mining project that acquired customary land, resulting in 570 families being affected and needing to be resettled. The resettlement process was delayed, which in turn affected their farming activities and overall well-being. The Human Rights Watch conducted a study on land-based investments in Serenje district (central Zambia), finding evidence of traditional authorities not consulting affected communities, as well as persistent failures by government agencies in providing oversight and enforcement of legal requirements (Human Rights Watch, 2017). Ultimately, this has led to the forced eviction of hundreds of individuals from their homes and lands with no compensation in most of the cases.

The example presented in Box 7.1 is a case of an LSLA that is considered to have avoided adverse displacement impacts, as international guidelines were followed and consultation and compensation were taken seriously. Nevertheless, the case presents a situation where investment has displaced several households and contributed to land scarcity in the area.

Perception on Land Tenure Security

According to a survey on tenure security, 27 percent of respondents in Zambia felt insecure about their tenure rights in 2018. Perceived tenure insecurity is higher in urban areas (Prindex, 2019).

The case discussed in Box 7.1 shows that irrespective of the actual changes to land tenure, areas in which investments take place can be marked by land scarcity despite the narrative of Zambia as having a lot of "idle land," and as a result perceived land tenure security degrades further in the presence of LSLAs.

Box 7.1 The case of Amatheon farm's impact on displacement & perceived tenure security

Amatheon Agri Zambia Ltd (heretofore Amatheon) is a large-scale farm operation based in the Big Concession farm block of Mumbwa district in central Zambia. The firm is a subsidiary of Amatheon Agri Holding N.V., an agribusiness and food company based in Berlin, Germany. Amatheon has acquired a leasehold of more than 40,000 hectares of state land, of which approximately 7000 hectares is operational at this time. The company started working in Zambia in 2012. It produces maize, soya, wheat, and groundnuts and owns 1300 heads of cattle. Amatheon employed 210 permanent workers at the time of writing. The number of casual workers who are employed for a maximum of six months varies between 300 and 600, depending on the season.

Besides its core farming business, Amatheon has set up an outgrower program with more than 12,000 registered smallholder farmers. The program started in 2015 in Mumbwa district and was co-financed by the Deutsche Investitions- und Entwicklungsgesellschaft (DEG). The program expanded to the district of Chibombo in 2017 when USAID joined the project. Amatheon uses the outgrower program as an instrument to increase its trading volume while also aiming to achieve a positive social impact. The company has established a network of rural trading depots in the two districts where farmers are able to purchase inputs and sell their crops and livestock. In addition, Amatheon, through its field officers and farmer coordinators, provides training to farmers on a number of topics, such as conservation agriculture, business and financial literacy, and post-harvest handling techniques. With the exception of a small agricultural finance component, Amatheon does not use contracts in its outgrower scheme. This implies that farmers can freely choose whom to sell their output to. However, it also means that Amatheon does not provide a purchase guarantee for farmers without input loans. While Amatheon purchased almost 11,000 tons of grain from 4045 farmers in 2016, the company only purchased 500 tons from 238 farmers in 2017. The reason for the low amount is a bumper harvest in the South African Development Community in 2017, which has led to a lack of demand and low market prices, especially for maize. According to company employees, Amatheon was not able to identify buyers to profitably resell farmers' produce. The outgrower program started off with a lot of ambition but has actually—so far—failed to meet expectations.

Several authors of this chapter have been involved in field research surrounding Amatheon's farm in Zambia, which was conducted between 2015 and 2018. Based on these insights, we share impacts of the Amatheon farm on displacements and perceived land tenure security.

Displacements: Amatheon refrained from purchasing customary land and only acquired state land. The land the company acquired was largely unin habited and in the hands of several (absentee) landlords. Although

(*continued*)

Box 7.1 (continued)

Amatheon sought to purchase only uninhabited land, many households still live(d) on the land. One reason was unclear land boundaries between customary and state land, making it difficult to identify truly uninhabited land. Amatheon decided to first develop tracts of land where only a few families lived. For the development of its first farm block and the construction of a dam, several families were resettled.[3] Amatheon used the Government of Zambia's Guidelines for the Compensation of Internally Displaced Persons and the International Finance Corporation's Handbook for Preparing Resettlement Action Plan Reports as a basis for drafting the resettlement plan. Reports show that families have been compensated and received titled land and brick houses (Chu et al., 2015; Herre, 2013); yet, in several instances, there have also been substantial delays in Amatheon meeting its agreements with the resettled families (Salverda, 2018).

The case of Amatheon also shows the challenges that investors might have in developing and safeguarding acquired land. Amatheon plans to develop a block of land close to the Kafue River. According to the farm manager, 99 families live on the land and it will be too expensive to compensate all of them.

Perceived land tenure security, land prices, land scarcity, and land disputes: In 2018, we conducted a household survey among smallholder farmers in Mumbwa and Chibombo district where Amatheon operates its outgrower scheme. Our survey data shows that 98 percent of the land farmed by the 797 interviewed households is customary land, while the remainder is titled land. We find that land prices have considerably increased since Amatheon came to the area. Prices have increased from an average of 713 Kwacha (approximately 54 USD in 2019) per ha between 2007 and 2012 to 984 Kwacha per ha (approximately 75 USD in 2019) between 2013 and 2018. Moreover, we find that since 2013, there has been a slight increase in the number of farmers acquiring new land.

(continued)

[3] It remains unclear how many households were resettled due to Amatheon's operations—in the first phase and afterward. According to company information, 11 households were affected in the first phase, whereas Salverda (2018) claims 20 families had been resettled or their plots had been "carved out" and another 7 families had been in the process of resettlement during Salverda's research. Carving out refers to households that remain on their land and are assisted by Amatheon to obtain title deeds—they are "carved out" from Amatheon's land (Joala et al., 2016). Concerning the development of the second phase, Amatheon's homepage publishes a media report that refers to "39 affected households and fields" according to Amatheon's Environmental Impact Statement (Amatheon Agri Holding N.V.—Commencing Second Phase of Agricultural Development in Zambia, 2014). According to Henley (2017), Amatheon displaced 43 families.

Box 7.1 (continued)

Interviewed households assess the land as scarce despite the narrative of Zambia having abundant land. One of the questions we asked in the household survey was, *"In your perception, do village headmen/authorities still have unallocated arable land that could be given to households in this area?"* Out of 796 respondents, 88 percent answered "No," highlighting a sense of land scarcity in the study area. Further, we found perceptions of land availability differed according to the distance to Amatheon's operations, with a greater proportion (91 percent) of respondents living within 25 kilometers of the Amatheon farming operation reporting they did not perceive there was still unallocated arable land in the area.[4]

During focus group discussions (FGDs) close to Amatheon's farming operations in 2015, smallholders expressed fear that land in the area is scarce and that their land is becoming more and more insecure due to the presence of Amatheon (three FGDs in October 2015). For instance, one participant stated that *"this issue of buying land everywhere, so many hectares, uh. It's giving us fear."* In the same FGD, one participant added, *"So what we saw is that our leaders, political leaders, were not ensuring that their people also have land"* (FGD men, October 14, 2015). In another FGD, someone, referring to Amatheon, claimed, *"Kaindu land is too sweet for them. They want to get everything"* (FGD outgrower, October 13, 2015). In a women's group, one participant worried, *"So we just get worried that this buying of land is just too much. Maybe they can even get land where we stay. Where will we go?"* (FGD women, October 14, 2015). Irrespective of the actual changes to land tenure, there is a strong perception of land scarcity even to the degree that land owned by smallholders on customary land is threatened.

There are two other factors besides Amatheon acquiring land that might have contributed to increased land pressure. First, there are other larger private investments within Mumbwa district. For instance, Chu et al. (2015) report of a sugar plantation in Chief Shakumbila's area that attempted to acquire 20,000 ha of customary land from Chief Kaindu for a game ranch, as well as several medium-scale commercial farms (over 20 ha). Second, during two FGDs in 2017 with Amatheon outgrowers and farmer coordinators, participants reported that farmers expand their area cultivated because the outgrower program has taught them to see farming as a business. *"They*

(continued)

[4] Measured from the farm offices.

Box 7.1 (continued)

now know the value of land. Everyone wants to cultivate more hectares so that they can have more produce. Because you know farming is a business" (FGD farmer coordinators, July 15, 2017). In another FGD, one participant stated, "*The land is now finished because of farming*" (FGD outgrowers, July 15, 2017), indicating there is no more land left in the area. In both FGDs, participants are of the opinion that farming as a business is a contributing reason why people are now fighting over land.

With an under-resourced State facing difficulties of keeping land registration, demarcation, and other cadastral services updated, land disputes are frequent in the case of Amatheon. With much of Amatheon's land not being developed yet, there have been a number of Zambians, often coming from elsewhere in the country, settling, or being resettled, on Amatheon land over the years. In one case, a smallholder, who Amatheon had actually compensated with a brick house and plot of land elsewhere in the area, remained on the land he had to vacate, and even started selling plots to others. Probably because Amatheon was not developing that land, he saw an opportunity and pretended to be the owner. Though the court has settled the case in favor of Amatheon, this and other cases are indicative of the tensions that may come with large land ownership in the area. To solve some of these tensions, or at least provide evidence to the surrounding residents where Amatheon's land officially starts, the company has been requested to demarcate the boundaries. However, marking land boundaries has led to new tensions, with the residents disputing either Amatheon's ownership or the location of the boundary between their and Amatheon's land. With only a small part of its total land developed so far, while the local population is simultaneously perceiving increasing land scarcity, Amatheon will probably continue to be confronted with land disputes. As disputes arise, they will have to try and resolve in court, with the assistance of the local government and/or traditional leadership, and/or by means of resettling the residents according to international standards.

Sources: Nolte and Subakanya (2016); Salverda (2018); primary data from own field research.

Pressure on the Zambian Land Tenure System

For years, customary land documentation in Zambia has remained elusive because of the belief that land ownership in this type of tenure system may not require documentation (Kaima, 2017). In more recent times, however, a number of traditional chiefs have started issuing certificates of ownership that are legally recognized and enforceable. This effort to strengthen (customary) tenure security is a clear sign that chiefs increasingly feel pressure on the land tenure system.

According to the 1995 Lands Act, the President, who is the custodian of all land, is required to consult local leaders in the event that land under customary tenure is to be alienated for large projects. The specific parties responsible for approving the conversion of customary land to leasehold title in Zambia are the traditional chiefs and local government leaders in the targeted locations. The Commissioner of Lands at the Ministry of Lands and Natural Resources is also meant to provide consent unless the process causes injustice, or is contrary to national interest or policy (Samboko et al., 2018). However, this approval process does not happen in practice. Therefore, the rise in large-scale commercial agriculture, in the context of weak land protection, is likely to push smallholders off their land, force them out of production, impact their right to food, and eventually compromise the country's peace and social cohesion (Elver, 2018).

A new draft of the land policy is intended to address these inherent challenges with land administration in Zambia, but it remains contested among traditional leaders who have rejected the policy due to the fear of losing power and a lack of engagement with them (House of Chiefs, 2018; Kapata, 2018). At the moment of writing (early 2020), the draft land policy is at a deadlock. Nonetheless, this again makes the pressure on the dual land tenure system evident. The national land policy has been adopted in May 2021 (Republic of Zambia, 2021). The deadlock following the draft land policy of 2017 was revived with a new draft land policy in 2020. The adopted national land policy is a much condensed version of earlier drafts. While civil society organizations welcome the adoption, recognize the inclusive approach in policy formulation, and the potential for more equitable land access, they also caution against the lack of clarity concerning customary land certificates and a lack of adequate provisions around informed consent in the face of large-scale land-based investments. The actual implementation of the policy is thus crucial (Land Portal, 2021; Zambia Land Alliance, 2021).

Outlook: LSLAs, Land Tenure Security, and Sustainable Development

This chapter has highlighted that the relationship between land tenure security and LSLAs is multi-faceted and complex. Land tenure security has immense importance as to *where* land acquisitions take place, and how they *impact* local communities, and the security of tenure may even *change* due to land acquisitions that may lead to displacements and put pressure on the land governance systems. This has been discussed on a global scale and in the case of Zambia.

A number of key messages emerge from our analysis. First, especially in countries with weak land tenure security, it is important to *not* welcome investors without a second thought. This is because investors might be tempted to take advantage of institutional weaknesses. Thus, scrutinizing investors and their projects before granting them access to land is crucial. Second, in contexts with stronger land tenure security, adverse impacts of LSLAs may be mitigated more easily. Accordingly, in contexts with weak land tenure security, it is even more crucial to take the needs of marginalized groups into account and make sure they do not lose access to land if LSLAs are implemented. Third, as LSLAs may have impacts on land tenure security, we need to understand that there is a window of opportunity for institutional change, and closely follow the implications of changes to the land governance systems that can be catalyzed by LSLAs.

From our analysis, we conclude that for LSLAs to be implemented in an economically, socially, and environmentally sustainable way, land tenure security plays a key role: both investors and affected communities need to rely on land tenure security, and only if tenure is secure does it makes sense for land users to use land in an environmentally sustainable way.

Obviously, the question of how to improve the security of tenure is not easy to answer and depends largely on the national context. There is no blueprint for improving tenure security across the globe. In the Zambian context, we currently see the process of reforming the land policy at a deadlock. A crucial step here is to ensure all relevant stakeholders are consulted when drafting the upcoming land policy. This could address apparent weaknesses of the land governance system resulting in very weak land tenure security, especially for marginalized groups such as women and those living on customary land.

Future research should focus on two main aspects. First, we need a much better understanding of the differential impacts of LSLAs given land tenure security can vary within and across people and areas. In particular, research should focus on the diversity of impacts on different social groups, including men and women, the youth, marginal groups, and better-off and poorer smallholders. Second, research should advance understanding of how land tenure security can be strengthened given the factors unique to specific local settings. This requires a clear understanding of all relevant stakeholders and their specific roles and interests. Comparative case studies can reveal variations within and across countries and identify potential diverse influence channels, such as the role of domestic and international civil society, community movements, and international donors and policy frameworks. Finally, better data is needed to understand LSLAs, where and how they occur, and to understand the national and global extent of LSLAs. A clearer picture of LSLAs is starting to emerge, but to develop robust, timely policies, a representative and more comprehensive analysis is needed.

References

Ahlerup, P., & Tengstam, S. (2015). *Do the land-poor gain from agricultural investments? Empirical evidence from Zambia using panel data* (Working Papers in Economics No. 624). Retrieved from University of Gothenburg, Department of Economics website: https://econpapers.repec.org/paper/hhs-gunwpe/0624.htm

Alden Wily, L. (2011). 'The law is to blame': The vulnerable status of common property rights in Sub-Saharan Africa. *Development and Change, 42*(3), 733–757. https://doi.org/10.1111/j.1467-7660.2011.01712.x

Ali, D., Deininger, K., & Harris, A.. (2016). *Large farm establishment, smallholder productivity, labor market participation, and resilience. Evidence from Ethiopia*

(7576). http://documents.worldbank.org/curated/en/650951468187134876/pdf/WPS7576.pdf

Ali, D., Deininger, K., & Harris, A. (2019). Does large farm establishment create benefits for neighboring smallholders? Evidence from Ethiopia. *Land Economics, 95*(1), 71–90.

Amatheon Agri Holding N.V.—Commencing Second Phase of Agricultural Development in Zambia. (2014). Retrieved October 14, 2019, from Amatheon Agri Holding N.V.—Commencing Second Phase of Agricultural Development in Zambia website: https://amatheon-agri.com/commencing-second-phase-of-agricultural-development-in-zambia/

Arezki, R., Deininger, K., & Selod, H. (2013). What drives the global 'land rush'? *The World Bank Economic Review.*

Baumgartner, P., von Braun, J., Abebaw, D., & Müller, M. (2015). Impacts of large-scale land investments on income, prices, and employment: Empirical analyses in Ethiopia. *World Development, 72*, 175–190.

Behrman, J., Meinzen-Dick, R., & Quisumbing, A. (2012). The gender implications of large-scale land deals. *Journal of Peasant Studies, 39*(1), 49–79.

Boone, C. (2014). *Property and political order in Africa: Land rights and the structure of politics.* Cambridge University Press.

Borras, S. M., & Franco, J. C. (2013). Global land grabbing and political reactions 'from below'. *Third World Quarterly, 34*(9), 1723–1747. https://doi.org/10.1080/01436597.2013.843845

Bottazzi, P., Crespo, D., Bangura, L. O., & Rist, S. (2018). Evaluating the livelihood impacts of a large-scale agricultural investment: Lessons from the case of a biofuel production company in northern Sierra Leone. *Land Use Policy, 73*, 128–137. https://doi.org/10.1016/j.landusepol.2017.12.016

Bottazzi, P., Goguen, A., & Rist, S. (2016). Conflicts of customary land tenure in rural Africa: Is large-scale land acquisition a driver of 'institutional innovation'? *The Journal of Peasant Studies, 43*(5), 971–988. https://doi.org/10.1080/03066150.2015.1119119

Chu, J., & Phiri, D. (2015). *Large-scale land acquisitions in Zambia* (Research Report No. 50; p. 40). Institute for Poverty, Land and Agrarian Studies.

Chu, J., Young, K., & Phiri, D. (2015). Large-scale land acquisitions, displacement and resettlement in Zambia. *PLAAS Policy Brief, 41.*

Chu, J. M. (2013). *Creating a Zambian breadbasket: 'Land grabs' and foreign investments in agriculture in Mkushi District, Zambia* (No. Working Paper 33; p. 27). Rotterdam: The Land Deal Politics Initiative (LDPI).

Cotula, L. (2011). *Land tenure issues in agricultural investment.* Rome: FAO.

Cotula, L. (2013). *The great African land grab? Agricultural investments and the global food system*. Zed Books.

Deininger, K., & Xia, F. (2016). Quantifying spillover effects from large land-based investment: The case of mozambique. *World Development, 87*, 227–241. https://doi.org/10.1016/j.worlddev.2016.06.016

Elver, H. (2018). *Report of the Special Rapporteur on the Right to Food on her Mission to Zambia.*

German, L., Cotula, L., Gibson, K., Locke, A., Bonanno, A., & Quan, J. (2018). *Land governance and inclusive business in agriculture: Advancing the debate*, 41.

German, L., Schoneveld, G. C., & Gumbo, D. (2011). The local social and environmental impacts of smallholder-based biofuel investments in Zambia. *Ecology and Society, 16*(4). https://doi.org/10.5751/ES-04280-160412

German, L., Schoneveld, G. C., & Mwangi, E. (2013). Contemporary processes of large-scale land acquisition in Sub-Saharan Africa: Legal deficiency or elite capture of the rule of law? *World Development, 48*, 1–18.

Henley, G. (2017). *Review of social issues for large-scale land investment in Zambia* (Working Paper No. 2017/42; p. 15). United Nations University World Institute for Development Economics Research.

Herre, R. (2013). *Agribusiness-Expansion, Land Grabbing und die Rolle europäischer privater und öffentlicher Gelder in Sambia: Eine Bewertung basierend auf dem Recht auf Nahrung* (R. van Ess, Trans.). Köln: FIAN Deutschland e.V.

Herrmann, R. T. (2017). Large-scale agricultural investments and smallholder welfare: A comparison of wage labor and outgrower channels in Tanzania. *World Development, 90*, 294–310. https://doi.org/10.1016/j.worlddev.2016.10.007

Honig, L., & Mulenga, B. P. (2015). *The status of customary land and the future of smallholder farmers under the current land administration system in Zambia*. http://www.iapri.org.zm/images/WorkingPapers/wp101.pdf

House of Chiefs. (2018). *Report of the First Ever Chiefs Indaba on the Draft National Land Policy*. Lusaka, Zambia.

Human Rights Watch. (2017). *'Forced to leave': Commercial farming and displacement in Zambia*. Human Rights Watch.

Jayne, T. S., Chamberlin, J., Traub, L., Sitko, N., Muyanga, M., Yeboah, F. K., Kachule, R., et al. (2016). Africa's changing farm size distribution patterns: The rise of medium-scale farms. *Agricultural Economics, 47*(S1), 197–214. https://doi.org/10.1111/agec.12308

Jayne, T. S., Chapoto, A., Sitko, N., Nkonde, C., Muyanga, M., & Chamberlin, J. (2014). Is the scramble for land in Africa foreclosing a smallholder agricultural expansion strategy? *Journal of International Affairs, 67*(2), 35–53.

Joala, R., Zamchiya, P., Ntauazi, C., Musole, P., & Katebe, C. (2016). *Changing agro-food systems: The impact of big agro-investors on food rights. Case studies in Mozambique and Zambia.* http://repository.uwc.ac.za/xmlui/handle/10566/4342

Johansson, E. L., Fader, M., Seaquist, J. W., & Nicholas, K. A. (2016). Green and blue water demand from large-scale land acquisitions in Africa. *Proceedings of the National Academy of Sciences, 113*(41), 11471–11476. https://doi.org/10.1073/pnas.1524741113

Kaima, R. (2017). *The legality of customary land documents within the existing formal legal system in Zambia.* Nyimba District Land Alliance.

Kapata, J. (2018). *Ministerial Statement. The Outcome of the National Validation Meeting for the Draft National Land Policy.* http://www.parliament.gov.zm/sites/default/files/images/publication_docs/MINISTERIAL%20STATEMENT%20BY%20THE%20MINISTER%20OF%20LANDS.pdf

Kelly, A., Bedford, C., Lawry, S., Musengezi, J., Naughton-Treves, L., Sunderlin, W., Veit, P., Masuda, Y. J., Barthel, K., Robinson, B. E., Childress, M., Game, E. T., & Ginsburg, C. (2017). *Mapping the global governance context for land tenure security.* Presented at the 2017 World Bank Conference on Land and Poverty.

Khadjavi, M., Sipangule, K., & Thiele, R. (2019). Social Capital and Large-Scale Agricultural Investments: An Experimental Investigation. *The Economic Journal.* https://doi.org/10.1093/ej/ueaa050.

Land Matrix. (2019). Land Matrix. Retrieved March 6, 2019, from Global Observatory website: https://landmatrix.org/global/

Land Portal (2021). Zambia Launches National Land Policy. https://www.landportal.org/node/99906

Lavers, T. (2012). 'Land grab' as development strategy? The political economy of agricultural investment in Ethiopia. *The Journal of Peasant Studies, 39*(1), 105–132. https://doi.org/10.1080/03066150.2011.652091

Lay, J., & Nolte, K. (2018). Determinants of foreign land acquisitions in low- and middle-income countries. *Journal of Economic Geography, 18*(1). https://doi.org/10.1093/jeg/lbx011

Lay, J., Anseeuw, W., Eckert, S., Flachsbarth, I., Kubitza, C., Nolte, K., & Giger, M. (2021a). Taking stock of the global land rush: Impacts and risks. Analytical Report III. Bern Open Publishing.

Lay, J., Nolte, K., & Sipangule, K. (2021b). Large-scale farms in Zambia: Locational patterns and spillovers to smallholder agriculture. World Development, 140, 105277. https://doi.org/10.1016/j.worlddev.2020.105277

Lund, C. (2008). *Local politics and the dynamics of property in Africa*. Cambridge University Press.

Matenga, C. R. (2017). Outgrowers and livelihoods: The case of Magobbo smallholder block farming in Mazabuka District in Zambia. *Journal of Southern African Studies, 43*(3), 551–566. https://doi.org/10.1080/03057070.2016.1211402

Matenga, C. R., & Hichaambwa, M. (2017). Impacts of land and agricultural commercialisation on local livelihoods in Zambia: Evidence from three models. *The Journal of Peasant Studies, 44*(3), 574–593. https://doi.org/10.1080/03066150.2016.1276449

Milimo, J. T., Kalyalya, J. H., Machina, H., & Hamweene, T. (2011). *Social impacts of land commercialization in Zambia: A case study of Macha mission land in Choma district*. International Land Coalition.

Mujenja, F., & Wonani, C. (2012). *Long-term outcomes of agricultural investments: Lessons from Zambia*. IIED.

Neef, A., & Singer, J. (2015). Development-induced displacement in Asia: Conflicts, risks, and resilience. *Development in Practice, 25*(5), 601–611. https://doi.org/10.1080/09614524.2015.1052374

Nhandu, T. (2017). *Baseline Report Zambia*. Transparency International Zambia. Land and Corruption in Africa Project (DFID).

Nolte, K. (2014). Large-scale agricultural investments under poor land governance in Zambia. *Land Use Policy*. https://doi.org/10.1016/j.landusepol.2014.01.014

Nolte, K., Chamberlain, W., & Giger, M. (2016). *International land deals for agriculture. Fresh insights from the land matrix: Analytical Report II*. https://doi.org/10.7892/boris.85304

Nolte, K., & Sipangule, K. (2017). *Land use competition in Sub-Saharan Africa's rural areas*, 4.

Nolte, K., & Subakanya, M. (2016). Relationship between large-scale agricultural investors and local communities: Lessons from two investments in Zambia. *IAPRI Policy Brief, 79*. http://www.iapri.org.zm/images/PolicyBriefs/ps_79.pdf

Nolte, K., & Väth, S. J. (2015). Interplay of land governance and large-scale agricultural investment: Evidence from Ghana and Kenya. *The Journal of Modern African Studies, 53*(1), 69–92. https://doi.org/10.1017/S0022278X14000688

Prindex. (2019). Zambia. Retrieved April 5, 2019, from https://www.prindex.net/data/zambia/

Raimondi, V., & Scoppola, M. (2018). Foreign land acquisitions and institutional distance. *Land Economics, 94*(4), 517–540.

Republic of Zambia (2021). National Lands Policy, May 2021. https://pmrc-zambia.com/wp-content/uploads/2021/05/National-Lands-Policy.pdf

Salverda, T. (2018). Facing criticism: An analysis of (land-based) corporate responses to the large-scale land acquisition countermovement. *The Journal of Peasant Studies, 0*(0), 1–18. https://doi.org/10.1080/03066150.2018.1439930

Sambo, P. T., Haywood, C., Wardell, D. A., Kibugi, R., & Segger, M.-C. C. (2015). *Enabling legal frameworks for sustainable land use investments in Zambia* (Occasional Paper No. 141; p. 80). Center for International Forestry Research.

Samboko, P. C., Kabisa, M., & Henley, G. (2018). Constraints to biofuel feedstock production expansion in Zambia. *Development Southern Africa*, 1–15. https://doi.org/10.1080/0376835X.2018.1508988

Schüpbach, J. M. (2014). *Foreign direct investment in agriculture: The impact of outgrower schemes and large-scale farm employment on economic well-being in Zambia* (ETH Zurich). https://doi.org/10.3929/ethz-a-010401958

Shete, M., Rutten, M., Schoneveld, G. C., & Zewude, E. (2016). Land-use changes by large-scale plantations and their effects on soil organic carbon, micronutrients and bulk density: Empirical evidence from Ethiopia. *Agriculture and Human Values, 33*(3), 689–704. https://doi.org/10.1007/s10460-015-9664-1

Sitko, N. J., & Chamberlin, J. (2016). The geography of Zambia's customary land: Assessing the prospects for smallholder development. *Land Use Policy, 55*, 49–60. https://doi.org/10.1016/j.landusepol.2016.03.026

Tagliarino, N. K. (2014). *Zambia: Conversions of customary land to leasehold title.* Focus on Land in Africa.

World Bank. (2010). *Rising global interest in farmland: Can it yield sustainable and equitable benefits?* World Bank.

Zaehringer, J. G., Wambugu, G., Kiteme, B., & Eckert, S. (2018). How do large-scale agricultural investments affect land use and the environment on the western slopes of Mount Kenya? Empirical evidence based on small-scale farmers' perceptions and remote sensing. *Journal of Environmental Management, 213*, 79–89. https://doi.org/10.1016/j.jenvman.2018.02.019

Zambia Land Alliance. (2021). Civil Society Welcomes Adoption of National Lands Policy. https://www.facebook.com/ZambiaLandAlliance/posts/pfbid0VvbojypZfbC3SCyS8eHQ9wDqh9pULDHA3d941U7a8Fgg2Gt-ny8Tr5bSdPG9KGYcgl

8

Tenure Security in War-Affected Scenarios: Challenges and Opportunities for Sustainability

Jon Unruh and Mourad Shalaby

Land Tenure Security and Armed Conflict: Before, During, After

Land tenure insecurity has a deeply intertwined and fraught relationship with armed conflict. Because the exercise of both land rights and civil conflict are spatial endeavors and are fundamentally about the relationships between people(s) with regard to land and territory, the interaction between the two is complex, nuanced, highly variable, and not easily separated. This is particularly the case when extreme and sufficiently widespread tenure insecurity in society is a fundamental reason for conflict to begin with. There are a variety of pathways from acute tenure

Portions of this chapter have previously appeared in Unruh (2002, 2004, 2011).

J. Unruh (✉)
McGill University, Montreal, QC, Canada
e-mail: Jon.unruh@mcgill.ca

M. Shalaby
The Global Environment Facility, Washington, DC, USA

M. B. Holland et al. (eds.), *Land Tenure Security and Sustainable Development*,
https://doi.org/10.1007/978-3-030-81881-4_8

insecurity to conflict, and from conflict to forms of intensified tenure insecurity. However, there are also opportunities for tenure security improvements subsequent to armed conflict to make real contributions to sustainability and durable peace. This chapter reviews some of these pathways and uses the case of Afghanistan for illustration.

While there are a variety of factors that can be part of a tenure insecurity contribution to periods of armed conflict (resource scarcity; poor land access; governance, rule of law, and political problems; identity; geography; history; ethnicity; and grievance), many countries are able to establish laws and institutions broadly seen as legitimate and fair in order to manage these. However, countries affected or threatened by such factors that also lack the political and institutional capacity to resolve the resulting land rights problems can find that their existing civil institutions cannot endure the stresses of large-scale unresolved land problems in society. This is especially the case where rural land rights are a fundamental unresolved problem in society. In such scenarios, tenure insecurity for significant segments of a national population can lead to an accumulation of confrontational ways to dealing with land rights problems which emerge from an increasingly divided society. The result is a build-up of competition, inequity, confrontation, grievance, resentment, and animosity, with no legitimate, fair way to manage all of these through a country's legal and institutional systems. In these cases, alternative informal institutions and approaches (such as resistance, insurgent, warlord, or mafia forms of land tenure, or extremist approaches involving land rights) can emerge from the absence of effective, legal institutions. These alternatives are able to operate within the fluidity, confrontation, and grievances of land crisis-ridden settings. Such crisis-based alternative informal institutions, which often belong to specific segments within a population, usually do not function in a fair and equitable manner in the context of broader society, and so ideally need to be replaced or reworked. But because such crisis situations are very different from land tenure situations in stable, adequately functioning, peaceful settings, solutions to such situations are also different. What may work well in stable, peaceful settings have proven extremely difficult to implement and operate in societies affected by pervasive, severe, unresolved land rights problems. In such difficult contexts, different interventions are needed in order to be able to (1) work within a conflict-prone setting; (2) meet short-term land rights security needs; (3) use land rights as a tool in recovery or improvement; and (4) transition to more stable and conventional land rights arrangements. This chapter considers the role of large-scale insecure land rights situations with regard to how these are both problems for conflict-affected countries but with significant lessons learned for sustainability. Subsequent to a review of tenure insecurity as a contributor to armed conflict, and the forms of tenure insecurity that then emerge from conflicts, the chapter looks at certain opportunities for post-war land tenure security in recovery and sustainability and concludes with a look at Afghanistan as an example of tenure insecurity as a contributor to armed conflict and as a lesson for the management of tenure issues subsequent to conflict.

From Tenure Insecurity to Armed Conflict

The importance of land rights issues as a contributor to civil conflict is reflected in the significant role that agrarian reform has played in many insurgent and revolutionary agendas. While there are several pathways from land rights problems to widespread tenure insecurity to armed conflict, they usually coalesce around pre-conflict perceptions of the 'unjustness' in the way the state deals with land rights for specific segments of a population. Left unattended, this then constitutes an important aggregate force in the deterioration of tenure security and the reduction of state legitimacy prior to the onset of conflict. Such perceptions can range from simple disappointment in, or distrust of the state and its ability, willingness, or bias in handling land issues, to the perception of the state as the enemy. The latter can be especially powerful if there are land-related grievances against the state brought on by specific issues that fuel tenure insecurity: land alienation and discrimination, legalized forms of eviction, land confiscations and speculation, crowding, corruption in court proceedings and court access, state intervention in agricultural production, dislocating agricultural and/or population programs, and heavy-handed approaches to enforcement of state decisions and prescriptions regarding land issues. Such dysfunctional statutory land tenure systems can be rife with micro-level generic disputes that do not get resolved and are often highly discriminatory. At times they can become, as in the Balkans, formal policy in support of ethnic cleansing (Toal & Dahlman, 2011). In Liberia prior to the war, the statutory tenure system generated an accumulation of rural underclass land-related grievances that resulted in a crisis of agrarian institutions, while at the same time poor governance precluded the peaceful derivation of legitimate alternatives (Sawyer, 2005). The result was the production of deep animosities regarding land that were a primary cause of the war (Richards, 2005). In El Salvador, grievances toward the landed elite and the state were at the core of the country's problems since the colonial era, and a primary cause of the conflict in the 1980s (de Bremond, 2007). This was also the case in Zimbabwe's liberation war due to land expropriations by the Rhodesian state (Mutasa, 2015), and in Mozambique's RENAMO war and Ethiopia's Derg war as a result of government villagization programs (Norfolk & Liversage, 2003; Wubne, 1991). Variants of such conditions also prevail for problems in southern Mexico, and in the way the land issue has been handled over the course of the conflict between the Palestinians and the Israelis (Unruh, 2002; Cohen, 1993). Such perceived injustices resulting in widespread tenure insecurity can become especially problematic if they merge with other issues involving the state not necessarily related to land, serving to further decrease the state's legitimacy. This was a fundamental part of the situation in Somalia, where disputes over access to grazing and water resources quickly merged with a history of perceived wrongs done by the state to certain clans and sub-clans on issues not directly about land (Besteman & Cassanelli, 1996). Animosities toward the state that become tied to historical events also have played a fundamental role in perceptions about who has legitimate access to what lands in the Balkans (Toal & Dahlman, 2011). In some scenarios, the accumulation of land-related grievances, combined with the lack of legitimate and workable alternatives, can lead to a search for order. Such was the case with the eventual emergence of Shari'a courts in Somalia, and, arguably, the emergence of the Taliban in Afghanistan. Both were able to field their own mechanisms of enforcement for a variety of institutions, including those managing land rights (Unruh, 2002).

For many who find themselves in worsening tenure security scenarios, identity can be, or can quickly become, intricately bound up in land occupation, access, or perceived rights to specific lands in very powerful ways. The existence of ethnic, religious, geographic, or other identities to which primary attachments persist can be based on connections to land, home area, or territory, with worsening tenure security in such a context then contributing significantly to animosity, and the development of narratives of grievance (Abdul-Jalil & Unruh, 2013). With armed conflict developing in such a context, some groups will seize the opportunity to advance the goals of substate self-determination, especially with regard to land and territorial rights. And as the identities of those involved in narratives of grievance develop to take on significant enmity with opposing groups (including the state), approaches to land issues will reflect this and can become a prominent feature in the conflict and subsequent peace process. In such a scenario approaches to land employed by one group in a conflict can be purposefully rejected by another. The difference between Palestinian and Israeli approaches to land and land tenure is in a number of ways grounded in identity. Identity for Palestinians especially has developed to a significant degree to mean opposition to Israel, Israelis, and Israel's approach to land administration (Unruh, 2002). Sri Lanka provides an example where land rights problems attached to identity were a primary cause of what became a 30-year war (Fonseka & Raheem, 2011; CPA, 2016). In Sri Lanka, long-standing grievances, discontent, and perceptions of government discrimination in land rights hardened along ethnic lines prior to the conflict (Yusuf, 2017; Rajasingham-Senanayake, 2005). Even years after the war ended in 2009, '[l]and is a key issue for reconciliation in Sri Lanka' (CPA, 2016).

From Conflict to Tenure Insecurity

The ongoing disintegration of land rights institutions during armed conflict, and yet the importance of land, homeland, and territory to the cause and conduct of conflict present particular dilemmas once hostilities cease. Like the complex histories involving land and territory that lead to conflict scenarios, the post-war re-establishment of ownership, use, and access rights will likewise be complicated and problematic. Left unattended, land and property issues can provide significant potential for renewed confrontation (Vines, 1996; Percival & Homer-Dixon, 1995; Crocker & Hampson, 1996). An end to armed conflict, especially prolonged civil conflict, creates a situation whereby a significant proportion of the affected population will begin to claim, re-claim, or access lands and land-based resources. The result is that land rights issues can be thrust to centerstage over large geographic areas in a short period of time for considerable numbers of people.

Significant change in tenure rights and security can occur during armed conflict and it is the outcome of these changes which are most operative at the close of a war. While such change can build upon prewar tenure problems, they nonetheless act to thrust the post-war land rights situation in new directions. Civil conflict necessarily results in a reduction in the power and penetration of state law, with the overall effect spatially variable. Early in a war, the state's land and property administration institutions in affected areas of the country can be damaged, destroyed, or rendered crippled or inoperable, and rules unenforceable—particularly if they were weak and/or linked to prewar grievances to begin with. This comes about due to general human insecurity; areas occupied by opposition groups or populations sympathetic to them; diversion of resources; departure of personnel who worked in service provision; and the destruction of the physical components of the lands system such as buildings, survey and demarcation equipment, boundary markers, local registries, and other records. In East Timor, the land and property buildings were among the first destroyed by militia activity along with most property rights records (Marquardt et al., 2002). As civil conflict grew in Somalia in the early 1990s, a reduced state capacity contributed to certain areas of the country being claimed by nomadic pastoralists under clan transient-access rights arrangements, by small-scale agriculturalists, by large-scale land interests accessing lands through the instruments of the crumbling state, and by heavily armed interests seeking access and control over lands by force (Besteman & Cassanelli, 1996). As well, forms of land tenure may be created which are directly connected to the opposition or insurgency which is made legitimate by direct military occupation and military strength (e.g., Vines, 1996).

Apart from the destruction of state capacity in land rights administration, armed conflict alters the system of rights and obligations in human relationships about land, with the result being that accepted and established rights arrangements can be at the forefront of change during conflict. The social and spatial repercussions of violence, dislocation, property destruction, battlefield victory and loss, and food insecurity, together with the breakdown of land-related norms, significantly alter ongoing relationships between people(s), land uses, production systems, and population patterns. In essence, armed conflict and its repercussions reconfigure the network of social relations upon which land tenure systems depend. For example, physical separation of people from established home areas and ways of land use and tenure due to wartime dislocation can be the first and most dramatic step toward the development of a changed approach to land rights. Physical separation from one's land changes, terminates, or puts on hold prevailing rights and obligations among people regarding land, especially where actual occupation or social position forms the basis or a significant aspect of claim. The result can be landholders abandoning the features of tenure systems because disputes and the lack of legitimate ways to resolve them have made such features unworkable. Or they believe there is little point in adhering to land rights rules that others are not following. And because those dislocated by war often develop or deepen political awareness while away from home areas, land problems in a post-war phase can easily be placed within the animosities of the larger political dynamic, further complicating the recovery of tenure systems (Alexander, 1992).

Claims to land and hence tenure security for societies emerging from armed conflict usually experience a significant 'proving' or evidence problem—particularly for returning dislocated populations. The issue of proving rights to post-war land claims in a way that is legitimate to claimants, authority structures (state and customary), and potential counterclaims becomes quite important subsequent to conflict, with direct repercussions on tenure security. While it can be assumed that evidence for claims must have effective dispute resolution institutions in place in order to be effective and provide for some form of tenure security, this is actually not the case in many instances where land administrative capacity is lacking. Where effective, legitimate institutions are lacking, the use of certain forms of landscape-based evidence ('facts on the ground') can emerge and be particularly strong in post-war scenarios in order to attempt to maintain or enhance tenure security. Especially valuable are 'facts on the ground' that connect with both customary and statutory definitions of claim, such as concepts of 'occupation' (Unruh, 2006). Clearing land is widespread as a means of creating visible evidence of occupation and thus claim in situations where institutions for adjudication are lacking, weak, or one-sided. This practice is of great concern for sustainability. Deforestation as a form of creating 'facts on the ground' evidence is widespread because it is so effective. Thus, the more lacking or compromised local to national institutions are for adequately dealing with evidence (claim, dispute resolution), the greater the tenure security need will be to make a strong visible argument for claim, in order to preempt the likelihood of a counter-claim and therefore the need for an institution to resolve a dispute (Unruh, 2006).

Opportunities for Post-war Land Tenure Security and Sustainability

Despite the problems with tenure insecurity prior to, during, and after a war, there do exist opportunities for addressing tenure security and its contribution to sustainability subsequent to armed conflict. In general, these can be grouped into opportunities that pertain to statutory land rights systems, customary rights systems, and how these interact.

Tenure Security Opportunities in Statutory Tenure Systems

Within the statutory system of a country emerging from armed conflict, in many cases land-related laws must undergo some form of reform in order to address post-war land rights concerns that cause ongoing acute tenure insecurity. There are two reasons for this. First, as noted, problematic land laws often contribute to the onset of a crisis, and so need to be reformed. Second, even well-functioning and fair land laws are usually not able to handle the particular problems that a country in crisis presents, such that existing laws often need to be amended or put on hold, and new laws enacted. There are three primary reform responses to land problems connected to the statutory system: (1) broad national land policy reform, (2) legal actions aimed at specific problems, and (3) institutional reform. Land policy reform seeks a broad, national-level improvement in tenure security and includes a broad-based process of consultation with affected communities and sectors (villagers, ex-combatants, refugees, commercial interests, government, etc.). The process is usually supported by a consortium of donors together with a government who does not have the capacity to undertake such an endeavor itself. Land policy reform after conflicts is an involved process, needing a good deal of capacity building, coordination, political will, donor involvement, money, and often a good deal of time.

Legal actions aimed at specific tenure insecurity problems is a much quicker approach than land policy reform, and more easily achievable, albeit with less scope, and thus fewer people (nationally) are likely to experience enhanced tenure security. Specific legal actions which are able to attend to distinct land problems in a post-war context can be quite useful until a broader land policy reform can be considered. Examples of such legal actions include:

1. Legal decrees that focus on specific society-wide land issues and are quickly derived, disseminated, enforced, and then terminated when the objective is obtained. Decrees can be used to temporarily manage land speculation; evictions; confiscations; duress, coerced and bad faith sales and purchases; and to invalidate specific forms of claims that are proving destabilizing. Decrees and their effects are largely seen as temporary, to be replaced by more robust forms of law later.
2. Legal rulings that resolve specific but potentially volatile problems for certain post-war communities. Liberia's experience with the problem of adverse possession (uncontested occupation for a period of time resulting in legal ownership) dealt with the question of whether or not the wartime and post-war periods should count as part of the period of 'uncontested occupation' needed for ownership claims via adverse

possession. This affected large numbers of squatters in long-term occupation situations, but also returning commercial interests who had title to the same lands. In such a situation, if there is not a clear legal ruling on the issue, then powerful interests can seek to violently evict squatters who are claiming, or may be about to claim, ownership under adverse possession (Unruh, 2009).

3. Rendering legal decisions that affect or resolve an entire category of land and property claims and/or dispute problems. Both Liberia and Mozambique have had positive experiences with this strategy. The Sirleaf administration in Liberia cancelled all of the forestry concessions as a legal decision due to pervasive fraudulent acquisition and the profound tenure insecurity it caused (GOL, 2006). And Mozambique dealt with whole categories of problematic land claims issued before and after its war, involving (1) whether or not Portuguese colonists or their descendants would be able to return to claim lands, (2) the need for concession holders to reapply under new rules that included more adequate interaction with local communities, and (3) the cancellation of certain categories of concessions due to fraudulent acquisition (Norfolk & Liversage, 2003).

Institutional reform attends to the issue of violence being a too-ready alternative with which to pursue land (and other) issues because post-war state institutions are crippled, corrupt, not legitimate, or nonexistent. In such a situation, working to purposefully include customary institutions which are able to garner legitimacy from a local population into the recovering statutory legal system can be a positive step. At the same time, providing forms of state legitimacy to certain customary institutions can be a shortcut to setting up workable institutions (OAU-IBAR, 1999). Ethiopia has had particular success with this approach in its restive regions (Sugule & Walker, 1998).

Tenure Security Opportunities in Customary Tenure Systems

The practical reality in many situations of low government capacity in land administration after a crisis such as armed conflict will be that customary and other forms of informal tenure are often the prevailing system(s) for the majority of the population, and where a great many attachments to tenure security are made. In such a context, opportunities for improvements in tenure security include what should be avoided. State actors and the international community present in post-conflict countries would do well to not insist on or attempt to re-impose debilitated or corrupt statutory law arrangements onto situations where customary law is re-emerging, as tenure insecurity for those within customary tenure systems would see a marked decline. At the same time, attempting (except in highly abusive circumstances) to downgrade customary law so as to promote statutory law in practice as a form of recovery would likely be counterproductive to tenure security. Prewar statutory land laws that are quickly reimposed after crises and in degraded institutional situations often have little ability to be enforced, are very open to corruption and abuse, and in many cases will have contributed in some fashion to the cause of the crisis.

While the use of certain 'facts on the ground' noted earlier (e.g., clearing to claim) that detract from sustainability can surge after armed conflicts in attempts to gain greater tenure security, other customary facts on the ground are much more sustainable and can be useful for enhancing tenure security. Purposefully planted economic or 'marker' trees are a good example of this due to the very clear connections made between rightful occupants and the land upon which trees are planted. Such trees are notable for their pervasive role as legitimate evidence for claim within customary systems, and their strong connection with formal legal concepts of long-term occupation or presence. Using purposefully planted trees for land claim, demarcation, and enhancing tenure security is widespread, particularly in low-capacity institutional environments such as after a war. Recognizing such customary forms of evidence for claims in reforming statutory laws and institutions has proven useful in a number of cases (e.g., Vogt et al., 2006; Unruh, 2002, 2008). In such situations, it is not tree planting after a war that is seen as valuable, but rather the recognition of the owner and tenure security value of older trees planted well before the war and their utility in re-attaching people to landscapes.

Societies emerging from armed conflict are usually fractured into units of lineage; extended family; ethnicity; tribe; religious; geographic; or experienced-based groups such as refugees, ex-combatants, female head of household, and squatters. In this context, with state capacity low, the emergence of multiple ways of using group legitimacy and authority to secure access to rural lands is common (Unruh, 2003; Galanter, 1981). In such a scenario, previous experiences with what is called 'forum shopping' can be useful. Forum shopping occurs when individuals and communities choose and negotiate with each other regarding which fora to go to in order to resolve land rights problems—disputes, claims, restitution, squatting, eviction, and so on (Galanter, 1981; Lund, 1996). Where such 'legal pluralism' is present, there can be a variety of authorities, rules, and institutions to choose from, including forms of customary law, informal wartime norms, remnants of formal law, hybrids of these, as well as the perceived legal capacities and institutions associated with humanitarian organizations, donors, and NGOs. While messy, forum shopping can offer room for maneuver or negotiability, potentially reducing tenure insecurity and violence in a degraded state administrative situation in the near term (Galanter, 1981; Lund, 1996). Allowing and managing such forum shopping in recovery scenarios can buy time while other forms of land rights recovery take shape, and can provide important ingredients for what works.

Lessons Learned for Afghanistan

The case of Afghanistan demonstrates how intertwined land rights and tenure security are with armed conflict and recovery, with important lessons for donors and governments engaged in sustainable peacebuilding. Afghanistan illustrates the current approach to peacebuilding and reconstruction by the international community after conflicts, which is to focus on the large-scale building blocks of recovery separately. However, such building blocks are largely isolated from each other in their planning, analysis, implementation, and measures for success with regard to contributing to overall peace and recovery. In Afghanistan's case, two of these building blocks—recovery of land rights and (re)construction of road infrastructure—are regarded separately as crucial to post-war recovery. However, how they interact has resulted in large-scale land grabbing, tenure insecurity, and obstacles to achieving durable peace (Unruh & Shalaby, 2012).

The reconstruction of road infrastructure, and in many cases its construction for the first time in war-torn countries, is thought to contribute, on its own, to peacebuilding and post-war recovery in very substantial ways. Road (re)construction is intended to facilitate trade and economic linkages, promote access to a wide variety of services, boost agricultural yields, bring rural areas into commercial interaction with the marketplace, provide security to rural communities, and contribute to the development of other sectors (JICA, 2006; JSCE, 2002; USAID, 2006, 2009). The realization of these benefits is understood to be crucial to economic and livelihood recovery and development, and hence the presumed winning of hearts and minds in unstable and volatile socio-political settings (JICA, 2004; Mockaitis, 2003; USAID, 2009). At the same time, the reconstitution of land and property rights systems in post-conflict scenarios is fundamental for the return of dislocated populations; restitution; improvements in tenure security; agricultural recovery and food security; broad economic recovery; and the ability to address volatile ethnic, tribal, and religious claims to lands (Bruch et al., 2009). The reconstitution of functioning land rights systems is also thought to contribute to the resolution of an array of problems recognized as important causes and catalysts for armed conflict—ethnic cleansing, evictions, retribution, inequality in land and property laws and legal systems that are non-inclusive or exploitive, and land-related grievances and animosities (Bruch et al., 2009; DW, 2005; Wiley, 2003).

Implemented in the same areas at the same time, these two priority components of peacebuilding do not however interact in planning, programming, implementation, or evaluation (JICA, 2006; JSCE, 2002; USAID, 2009). They do, however, interact quite robustly on the ground among recipient populations in an unplanned way, to produce very difficult outcomes, with some of these working significantly against sustainable recovery. A primary outcome of the interaction between road (re)construction and land tenure in the prevailing context in Afghanistan is a large surge in land grabbing—resulting in widespread and dramatic declines in tenure security. This occurs as the increases in land values brought on by road (re)construction occurs, within a context of a debilitated capacity of both customary and statutory tenure systems, increased ease of access to lands (via roads), flourishing corruption, and the absence of many landowners due to dislocation. Land grabbing in Afghanistan is so acute that it is thought to constitute a significant conflict-related flashpoint, able to push the country into renewed civil unrest (Batson, 2008; IWPR, 2008). Land grabbing by powerful interests, including government officials, militia commanders, former military commanders, and members of parliament is pervasive (Irvine, 2007). Of particular importance is the 3000 km ring road connecting Kabul, Herat, Kandahar, and back to Kabul (Fig. 8.1). When completed it is estimated that 60 percent of Afghans (approximately 17 million people) will live within 50 km of this road (USAID, 2009). However, of the nine provinces where the percentages of government-seized agricultural land are the highest (Reydon, 2007), all reside along the ring road. Six of these provinces have had between 80 and 90 percent of their agricultural area subjected to land grabbing and three have had over 100 percent of the land grabbed (Baghlan province, 110 percent; Kandahar, 111 percent; and Logar, 190 percent) (Reydon, 2007), indicating that land is being grabbed, and then grabbed again.

The resulting severe tenure insecurity brought on by land grabbing, along with the government's institutional deficit, then strongly dissuades people from engaging in any services connected with the state land tenure system that should be used to enhance tenure security—laws, titles, inheritance procedures, the official land market, the recording of transactions, surveying, demarcation, and so on. Instead, these are all seen as ways that the officialdom comes to know about the location, size, and potential value of lands, and importantly, which land can be most easily seized. This then encourages a search by local inhabitants for alternatives to state institutions for forms of security of their lands which are able to resist or act against land grabbing brought on by road reconstruction and

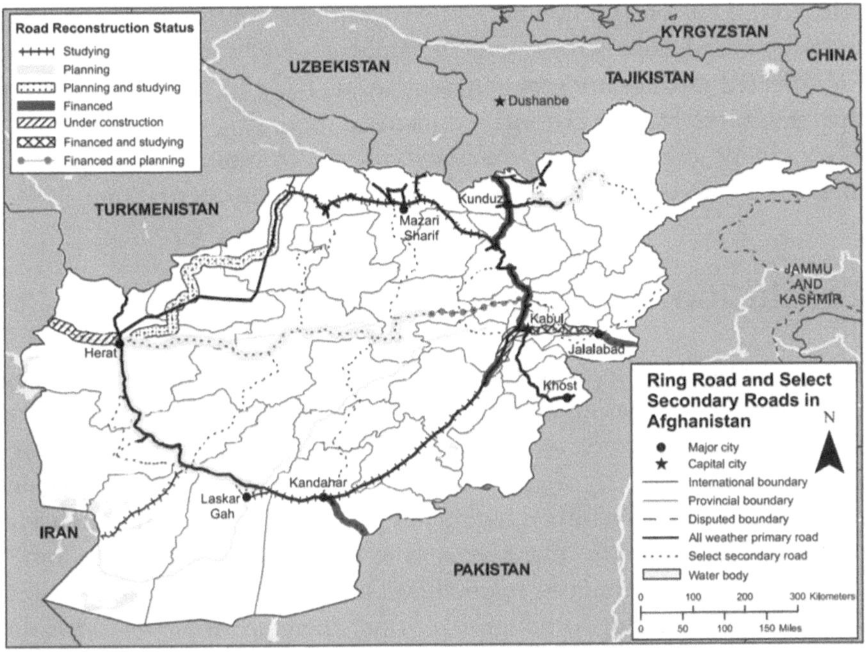

Fig. 8.1 Afghanistan's donor-funded ring road, as part of the country-wide road reconstruction effort. Source: Unruh and Shalaby (2012)

corruption. The Taliban are only too eager to provide such an alternative, especially as this often involves violent actions against state actors. The Taliban also provide disgruntled and disenfranchised villagers with weapons so they are able to react to land grabbing themselves (Sato, 2010). In fact, the Taliban have used widespread discontent with the land-grabbing problem specifically as a recruiting tool as resentment over the problem and desire for retribution and restitution of seized land and property grew after the war (Bowman, 2010).

What the Afghanistan case demonstrates for tenure security in recovery scenarios is that the sequencing of interventions is of utmost importance. In this case, the reconstruction of the land rights system—legal, technical, political—should have come before the implementation of road reconstruction efforts. Much greater attention could have been placed on capacity building for institutions, services, and rule of law, prior to road reconstruction efforts. An additional important lesson from the Afghanistan case is that, while it can be generally acknowledged that war-affected settings are very different from stable settings, exactly how this translates technically into implementation of the large building blocks in peacebuilding by the international community and governments needs to be thoroughly analyzed prior to implementation.

Conclusion

Tenure (in)security in countries threatened by, enduring, or recovering from civil conflict has great influence on the prospects for sustainability. War-torn, fragile, failed, and in-transition states have particular difficulty aligning tenure security with sustainability so that the two are mutually reinforcing. Instead, tenure insecurity aligns with a deterioration in livelihoods and land resource management, to draw communities into sometimes prolonged periods of fragmentation, narratives of grievance, and violent confrontation. Although daunting, there are approaches for the reorientation of tenure insecurity in such fraught settings toward greater security. And while this chapter mentions a just few of these, there are now a significant number of country experiences that are coming together to comprise best practices.

References

Abdul-Jalil, M. A., & Unruh, J. D. (2013). Land rights under stress in Darfur: A volatile dynamic of the conflict. *War and Society, 32*, 156–181.

Alexander, J. (1992). Things fall apart, the center can hold: Processes of postwar political change in Zimbabwe's rural areas. In T. Ranger & N. Bhehe (Eds.), *Zimbabwe's liberation war*. James Currey.

Batson, D. (2008). *Registering the human terrain: A valuation of cadastre*. Center for Strategic Intelligence Research, NDIC Press.

Besteman, C., & Cassanelli, L. (Eds.). (1996). *The struggle for land in southern Somalia: The war behind the war*. Westview Press.

Bowman, T. (2010). Karzai's brother tied to corrupt Afghan land deals. *National Public Radio*, 1 February. http://www.rawa.org/temp/runews/2010/02/01/karzai-brother-tied-to-corrupt-afghan-land-deals.html

Bruch, C., Jensen, D., Nakayama, M., Unruh, J., Gruby, R., & Wolfarth, R. (2009). Post-conflict peacebuilding and natural resources. *Yearbook of International Environmental Law, 19*, 58–96.

Centre for Policy Alternatives (CPA). (2016). *Land occupation in the Northern Province: A commentary on ground realities and recommendations for reform*. Centre for Policy Alternatives.

Cohen, S. (1993). The politics of planting: Israeli-Palestinian competition for control of land in the Jerusalem periphery. *Geography Research Paper No. 236*. University of Chicago Press.

Crocker, C., & Hampson, F. O. (1996). Making peace settlements work. *Foreign Policy, 104*, 54–71.

de Bremond, A. (2007). The politics of peace and resettlement through El Salvador's land transfer programme: Caught between the state and the market. *Third World Quarterly, 28*, 1537–1556.

Development Workshop (DW). (2005). Terra, urban land reform in postwar Angola: Research, advocacy and policy development. *Development Workshop, Angola Occasional Paper No. 5*. Luanda.

Fonseka, B., & Raheem, M. (2011). *Land in the Northern Province: Post-war politics, policy and practices*. Centre for Policy Alternatives.

Galanter, M. (1981). Justice in many rooms: Courts, private ordering and indigenous law. *Journal of Legal Pluralism, 19*, 1–47.

Government of Liberia (GOL). (2006). *Executive order #1*. Government of Liberia.

Institute for War and Peace Reporting (IWPR). (2008). *Afghan environmental health problems legacy of land-grab*. Institute for War and Peace Reporting.

Irvine, S. (2007). Powerful grab Afghanistan land. *BBC News*. http://news.bbc.co.uk/go/pr/fr/-/2/hi/south_asia/6981035.stm

Japan International Cooperation Agency (JICA). (2004). *Afghanistan assistance: From reconstruction to development*. Japan International Cooperation Agency (JICA).

Japan International Cooperation Agency (JICA). (2006). *The project formulation study on road maintenance and management sector in Afghanistan*. Japan International Cooperation Agency (JICA) and Katahira & Engineers International.

Japan Society of Civil Engineers (JSCE). (2002). *Afghanistan reconstruction vision*. Japan Society of Civil Engineers and Kabul University.

Lund, M. (1996). *Preventing violent conflicts: A strategy for preventive diplomacy*. United States Institute of Peace Press.

Marquardt, M., Unruh, J., & Heron, L. (2002). *Land policy and administration: Assessment of the current situation and future prospects in East Timor, final report*. US Agency for International Development and Government of East Timor.

Mockaitis, T. (2003). Winning hearts and minds in the 'war on terrorism'. *Small Wars & Insurgencies, 14*, 21–38.

Mutasa, C. (2015). A brief history of land in Zimbabwe: 1890-today. *Focus on Land in Africa*. Retrieved September 24, 2019, from http://www.focuson-land.com/download/55031523be8cd/

Norfolk, S., and Liversage, H. (2003). *Land reform and poverty alleviation in Mozambique*. Paper for the Southern African Regional Poverty Network, Human Sciences Research Council.

Organization of African Unity - Inter-African Bureau for Animal Resources (OAU/IBAR). (1999). *Border harmonization meeting—introduction and agenda*. Border Harmonization Meeting 19–21 August, 1999. Organization of African Unity—Inter-African Bureau for Animal Resources, Lodwar Kenya.

Percival, V., & Homer-Dixon, T. (1995). *Environmental scarcity and violent conflict: The case of South Africa*. Project on Environment, Population, and Security, University of Toronto.

Rajasingham-Senanayake, D. (2005). Sri Lanka and the violence of reconstruction. *Development, 48*, 111–120.

Reydon, B. P. (2007). Social embeddedness, institutions for rural land management and land grabbing, the cases of Brazil and Afghanistan. Unpublished manuscript.

Richards, P. (2005). To fight or to farm? agrarian dimensions of the Mano River Conflicts Liberia and Sierra Leone. *African Affairs, 104*, 571–590.

Sato, M. (2010). *Demobilization, reintegration and natural resources in Afghanistan: Afghanistan's new beginnings programme (ANBP)*. Paper presented at the Second International Symposium on Strengthening Post Conflict Security and Diplomacy: Policy Recommendations to Integrate Natural Resources, Infrastructure and Peacebuilding. University of Tokyo, 24–25 June.

Sawyer, A. (2005). *Beyond plunder: Toward democratic governance in Liberia*. Lynne Rienner Publishers.

Sugule, J., & Walker, R. (1998). *Changing pastoralism in the Ethiopian Somali National Regional State (Region 5)*. South East Rangelands Project, UNDP-Emergencies Unit for Ethiopia.

Toal, G., & Dahlman, C. (2011). *Bosnia remade: Ethnic cleansing and its reversal.* University Press.

United States Agency for International Development (USAID). (2006). *Data sheet: Afghanistan, agriculture program, economic growth, agriculture and trade pillar.* US Agency for International Development.

United States Agency for International Development (USAID). (2009). *Afghanistan: Infrastructure.* US Agency for International Development.

Unruh, J. (2002). Local land tenure in the peace process. *Peace Review, 14*, 337–342.

Unruh, J. (2003). Land tenure and legal pluralism in the peace process. *Peace and Change: A Journal of Peace Research, 28*, 352–376.

Unruh, J. (2004). *Land and property rights in the peace process.* Intractable Conflict Knowledge Base Project, Conflict Research Consortium. University of Colorado. http://www.beyondintractability.org/m/Land_tenure.jsp

Unruh, J. (2006). Land tenure and the 'evidence landscape' in developing countries. *Annals of the Association of American Geographers, 96*, 754–772.

Unruh, J. (2008). Catalyzing the socio-legal space for armed conflict: Land and legal pluralism in pre-war Liberia. *Journal of Legal Pluralism and Unofficial Law, 58*, 1–32.

Unruh, J. (2009). Land rights in postwar Liberia: The volatile part of the peace process. *Land Use Policy, 26*, 425–433.

Unruh, J. (2011). Land rights and peacebuilding: Challenges and responses. *International Journal of Peace Studies, 15*, 89–125.

Unruh, J., & Shalaby, M. (2012). A volatile interaction between peacebuilding priorities: Road infrastructure (re)construction and land rights in Afghanistan. *Progress in Development Studies, 12*, 47–61.

Vines, A. (1996). *RENAMO: From terrorism to democracy in Mozambique?* James Currey.

Vogt, N., Bahati, J., Unruh, J., Green, G., Banana, A., Gombya-Ssembajjwe, W., & Sweeny, S. (2006). Integrating remote sensing data and rapid appraisals for land-cover change analyses in Uganda. *Land Degradation and Development, 17*, 31–43.

Wiley, L. (2003). *Land rights in crisis: Restoring tenure security in Afghanistan.* Afghanistan Research and Evaluation Unit.

Wubne, M. (1991). Resettlement and villagizaiton. In T. Ofcansky & L. Berry (Eds.), *A country study: Ethiopia.* Library of Congress Federal Research Division.

Yusuf, S. (2017). Generating empathy, restoring dignity: The curious case of Sri Lanka. *South Asia Democratic Forum, Brussels, SADF Focus Report No. 30.*

9

Land Tenure Insecurity and Climate Adaptation: Socio-Environmental Realities in Colombia and Implications for Integrated Environmental Rights and Participatory Policy

Brianna Castro and Christina Kuntz

Our planet has warmed 1 degree Celsius since the nineteenth century. In 2018, the Intergovernmental Panel on Climate Change released a landmark special report demonstrating how the climate has changed and the future impacts that could be avoided if continued global warming is limited to 1.5 degrees Celsius. To do so requires reaching 'net zero' carbon dioxide (CO2) emissions by 2050, which could only be accomplished by overhauling the global economy. Even with warming limited to 1.5 degrees Celsius, 70–90% of coral reefs will be lost, 14% of the world population will experience extreme heat waves about every five years,

B. Castro (✉)
Harvard University, Cambridge, MA, USA
e-mail: briannacastro@g.harvard.edu

C. Kuntz
International Development Consultant, Cartagena, Colombia

M. B. Holland et al. (eds.), *Land Tenure Security and Sustainable Development*,
https://doi.org/10.1007/978-3-030-81881-4_9

droughts will be more frequent, and sea levels will continue to rise (IPCC, 2018). Global commitment to climate mitigation is evident in the United Nations Sustainable Development Goals, which include climate change mitigation and natural resource protection, the Paris Accords signed in 2016 agreeing to mitigate climate change through lowered greenhouse gas emissions, and national-level climate adaptation plans. However, progress toward these goals to date has been limited and CO2 emissions persist at unmitigated rates.

While cities, national governments, the private sector, and intergovernmental organizations work out whether and how to mitigate climate change through net-zero CO2 emissions, communities face more frequent and severe quick-onset disasters as well as gradual changes in seasons, temperatures, and rainfall. Global temperatures are rising, especially in the tropics, as is the incidence of extreme hot and extreme cold days, and coastal land is disappearing into the sea at rapid rates (IPCC, 2018). Multi-scalar climate adaptation in response to these impacts is already occurring through the adjustment of natural and human systems. In rural developing regions, adaptation to climate change most often involves changes in resource use. Unlike in more urbanized areas where planned adaptation prioritizes improved infrastructure or coordinated state programs, in the rural developing world households turn to adapting their homes and livelihoods on their own (Adger et al., 2003; Smit et al., 2000; Masuda et al., 2019). This local-level, autonomous adaptation requires new ways of using land in place, temporary mobility during environmental stress, or permanent migration (Fankhauser et al., 1999). Whether planned or autonomous, adaptation in the rural developing world is intrinsically connected to land tenure security, which can enable or impede people's capacity to adapt to shifting resource availability and climate conditions.

We take this insight—that adaptation in the rural developing world hinges on security of rights and access to land and resources—as our starting point for this chapter on climate adaptation and land tenure security. After discussing how climate change squeezes resources and the socioeconomic impacts of that pressure, we review research on land tenure security and climate change adaptation. We then examine the connection of complex land tenure history and climate stress in Colombia,

South America, one of the world's most biodiverse countries where complicated tenure has increased displacement, inequalities, conflict, and natural resource destruction, thus resulting in increased social vulnerability and impeded climate adaptation for rural populations. We use a protracted drought in the rural Montes de María region of Colombia to illustrate how households' adaptations are impacted by (mis)understandings of their rights to land and tenure laws in a country in a transitional era toward post-conflict. As land tenure realities and the effects of climate change vary drastically by location, studying specific cases in the rural, developing world can help policymakers better understand the relationships between these variables. Policy that is flexible, attuned to environmental rights, and is coupled with effective citizen participation models can help ensure sustainable implementation of land tenure programs and heightened climate adaptation planning for some of the most vulnerable communities.

Winners and Losers in the Climate Change Landscape

The changing climate significantly impacts land by causing it to be degraded or disappear. As land is disappearing and degrading, competition over freshwater access and land resources increases in tandem. Though the global community works to mitigate climate change impacts, its effects on land persist and are not experienced equally across the globe. Furthermore, the states that contribute most to human-caused climate change through emissions experience less of these land impacts than states that have contributed least to rising temperatures (Andrew et al., 2018).

Global sea levels have been on the rise for a century, but the rate of rising waters has escalated over the past few decades and is expected to accelerate dramatically in the decades to come (IPCC, 2018). The effects of seas rising at accelerated rates are intensified by increased settlement along coastlines over recent decades made possible through hardened infrastructure and engineered coastal protection through dikes, sea walls, and levees. These infrastructural protections, however, were built based on past estimates of rising seas rather than the accelerated pace of sea level rise due to climate change (IPCC, 2018). Coastal land and Pacific Islands are subsiding due to the sea level rise, and this land disappearance will increase at an exponential pace. Habitable land on the island of Nuatambu, for example, has decreased by over 50% since 2011 (Albert et al., 2016) and five Pacific Islands have already disappeared.[1] Up to 180 million people worldwide are directly at risk of the sea submerging the land where they live, and over 1 billion people live in low-elevation coastal zones that will experience the effects of proximate rising seas (Hoegh-Guldberg et al., 2018; Neumann et al., 2015). These effects include eroded coastlines and saltwater infusion, which changes vegetation and wildlife resources that communities depend on for their livelihoods, as well as inhibits agriculture in coastal areas. Land disappearing in the sea creates complicated land tenure dilemmas including, for example, whether or not those who owned the land that disappears will be compensated, if those without secure tenure who lose the land on which they lived have any recourse, and how to resettle communities on land projected to be uninhabitable due to sea level rise (IPCC, 2018).

While sea level rise impacts are experienced along coastlines, land degradation permeates the global landscape and concentrates in tropical and subtropical regions. Shifts in average temperatures, extreme hot and cold days, and precipitation add increasing pressure to existing land resources through shifts in soil, water availability, and biodiversity. Increased droughts and soil erosion cause land degradation and force changes in the use of arable land (Ahmed et al., 2016). Countries in the tropics and Southern Hemisphere subtropics expect the greatest impacts on economic growth caused by climate change. Net reductions in the harvests of maize, rice, wheat, and other cereal crops are expected and will be significantly worse in sub-Saharan Africa, Central and South America, and Southeast Asia (Hoegh-Guldberg et al., 2018).

In the face of these direct climate impacts, the international community demonstrates commitment to mitigating climate change in a

[1] https://theconversation.com/sea-level-rise-has-claimed-five-whole-islands-in-the-pacific-first-scientific-evidence-58511.

number of ways. As of April 2019, 185 parties committed to maintain the global average temperature increase to below 2 degrees Celsius through the Paris Agreement. The international commitment to increasing resilience and adaptive capacity, as well as mitigating climate change, is outlined in the United Nations Sustainable Development Goals, specifically SDG #13, and global finance flows to fund climate mitigation and adaptation increased 17% in the year after the Paris Agreement was reached (United Nations, 2018). This commitment continues at the national and subnational levels, where governments have created national, sub-national, and city-level climate adaptation plans outlining the necessary steps to adapt physical and social infrastructure to protect populations from climate hazards. These large-scale climate mitigation efforts, however, disproportionately shift how land is used in developing countries (Sunderlin et al., 2018).

Though the global community has made ambitious commitments to climate mitigation and adaptation, these commitments are outpaced by the rate at which the climate is changing. Catching up to the pace of climate change requires quicker economic restructuring to eliminate harmful greenhouse gas emissions, access to climate finance needs to be accelerated, and adaptation efforts rescaled to match climate hazards (United Nations, 2018; IPCC, 2018). While the global average forest loss is slowing down and increased financial resources are contributing to biodiversity protection, biodiversity loss and land degradation continue at disturbing rates and forest loss has spiked in key places. Trends of deforestation and emissions show that global commitments are not enough as greenhouse gas emissions reached record highs in 2017, after the Paris Accords were signed, and more than one-fifth of the Earth's total land area has been degraded since the turn of the century by human-induced processes such as cropland expansion and urbanization (United Nations, 2018).

These land effects of climate change have variable impacts globally. Those who are more socioeconomically vulnerable experience more harm as the climate changes (Morton, 2007; Thomas & Twyman, 2005). There is stark inequality in the way countries bear the brunt of climate impacts globally—where the majority of countries with the highest emissions are the least vulnerable to climate change impacts. At the same time, of the lowest greenhouse gas emission countries, the majority are extremely vulnerable to climate impacts (Althor et al., 2016). Developing countries are more likely to experience less habitable land, less hospitable conditions for human settlement, and less food and water availability (Morton, 2007). Areas most at risk of forest degradation, desertification, rising seas, encroaching salt water, and loss of biodiversity are found in developing countries and at the margins of developed countries such as tribal communities in coastal North America. The tropics and Southern Hemisphere subtropics expect the greatest impacts on economic growth resulting from climate change as well (Hoegh-Guldberg et al., 2018). Such losses in developing countries are rooted in climate change, combined with long histories of colonization and development (O'Brien & Leichenko, 2000; Paprocki, 2018).

The Tangled Web of Land Tenure Security and Adaptation

'Adaptation' to climate change (as used by the IPCC) describes the steps taken to decrease the climate hazard vulnerability of populations and infrastructure (2018). Though adaptation can occur at any scale (international, national, or local), our focus is on local-level adaptation that occurs either in anticipation of future climate risks or during climate stress. Adger et al. (2009) outline three moments when climate adaptation occurs—the first adaptation response occurs when an environmental condition necessitates an adaptation response. This would be when a farmer who relies on rain-fed agriculture waits to plant seeds, for example, because the seasonal rains have not begun when expected. The second moment occurs when the initial adaptation becomes insufficient and stops working. In this case, the same rural farmer has waited to plant their seeds but the rain season never arrives and planting is not feasible. If drought conditions continue into the following harvest season, they may adapt in additional ways, such as planting in a different location or planting without rain and bringing water from nearby freshwater sources. Typically, these second phase adaptations require greater resources in the form of financial investments, tools, or knowledge. A third moment, and the most extreme, is when the nature of the relationship between the human and their environment has substantially changed. This would mean that a farmer could no longer productively farm on that land and their land-use activity must shift entirely. Rural dwellers may adapt through either temporary or permanent migration of one or more household members after exhausting options for autonomous climate adaptations in place (Castro, 2019; Meze-Hausken, 2008; Stark & Bloom, 1985). In some cases, adaptations in place fail but populations remain 'trapped' in place amid climate stress due to socioeconomic factors (Black et al., 2013; Schewel, 2019).

In the climate change landscape, where communities in rural developing zones face rapid and unpredictable climate hazards, the long-term development goal should be to foster adaptable and flexible livelihoods to weather these pivotal thresholds of adaptation (Adger et al., 2009; Bardsley & Hugo, 2010). Local- and regional-level studies of adaptation in resource-dependent societies show considerable flexibility and resourcefulness to climate change (Mortimore & Adams, 2001; Reij & Waters-Bayer, 2014). Depending on the local-level context, communities rely on different mechanisms and perspectives in their adaptations. In Cameroon, rural villages coordinate their adaptations to environmental stress through local-level institutions that are cornerstones of rural life and land management in particular. These coordinated efforts improve villages' abilities to be more resilient to climate-related stress through forest protections, coordinating land use during crisis, and providing insurance through savings groups for future emergencies (Brown & Sonwa, 2015). In Ghana, social vulnerability and dependence on climate-sensitive occupations require adaptation among farmers. Frequent adaptations include crop diversification, engagement in non-farm secondary employment, rural-urban migration, and increasing the amount of land farmed (Dumenu & Obeng, 2016). Yet in the Rural Sahel, though farmers are aware of climate change, they often attribute issues of land use and livelihood change to economic, political, and social rather than climate factors. As such, their adaptations in land use and livelihoods due to climate stressors are attuned to non-climate, rather than climate, factors emphasizing the importance of framing climate change as an adaptation driver in adaptation programs and policies to ensure climate-friendly land-use adaptation responses (Mertz et al., 2009).

Climate change is a threat multiplier, however, which intensifies existing risks across the board in terms of human security and conflict in climate-vulnerable regions (Unruh & Abdul-Jalil, 2012), while the resulting droughts and floods add pressure to land and land use due to strategies employed in the early phases of local adaptation and mitigation. Early-stage adaptations include working land harder, depleting resources on the land itself (Morton, 2007). Local-level adaptation mechanisms such as those described here will face increasing pressures as climate impacts escalate and resource pressures can provoke conflict. The relationship between land tenure security and climate adaptation functions in a number of ways as rural dwellers adapt in place and later through adaptive migration. Shifts in land tenure can be expected as landowners farm more land to increase crop yields, and therefore acquire more land to do so. Climate change creates insecure tenure by making it more difficult for farmers to support and maintain the land currently in their possession during climate stress. The opposite is also true, depending on the context, where landowning farmers sell parts of their land for financial resources to cope with failed harvests. Farmers may also adapt by temporarily leaving their land with the expectation of return once the land has recovered, but others may occupy the land without permission during this period, resulting in tenure conflict over land claims after the climate crisis subsides. This buying, selling, and illicit acquisition of land in the context of climate stress creates vulnerability and the possibility of exploitation either among farmers with varying resources or by large landowners or multinational corporations interested in acquiring land.

When in situ adaptations fail, mobility is the most important adaptation to climate stress especially in response to variations in rainfall and droughts. Land tenure insecurity constrains families' either temporary or permanent mobility. During climate stress, farmers without formalized land tenure are likely to exhibit greater mobility in search of more hospitable farming environments. These mobile farmers without land tenure security may encroach and individuate communal lands or lands with questionable tenure (Morton, 2007). Such mobility without tenure increases social vulnerability and often pushes families to the most climate-vulnerable areas, those considered unfit for tenure and residence along land prone to landslides, above marsh water, or at the borders of nearby municipalities.

Complicated Land Tenure in Colombia (Past and Present)

Socio-political events throughout Colombian history have created lasting complexities in land tenure, especially in rural communities. In such communities, households may make climate-adaptation decisions based on their understanding of precarious rights to land and resources. Rural inhabitants in Colombia have a profound connection with, and dependence on, land in ways that urban populations do not. In rural settings, 'land' dictates people's livelihoods, production methods, consumption patterns, and interactions with the larger society. The control of land has been one of, if not *the* most, influential perpetrators of Colombia's ongoing armed conflict, which has lasted for over 60 years and has dramatically affected Colombia's people and natural resources. As the demand for agricultural products increased in the 1920s, so did the demand for land (del Pilar Lopez-Uribe & Sanchez Torres, 2018). However, most smallholder farmers did not have formal land titles, and agrarian reforms starting in the 1930s inadequately protected their land rights. All of these factors allowed for the privatization of rural land by larger landowners. As a result, groups of smallholder farmers organized themselves to form revolutionary movements that would later evolve into guerrilla groups in order to reclaim rural land on which they used to farm but no longer had access (Alfonso Sierra et al., 2011).

As these smallholder farmer-based groups gained power, paramilitary groups formed to protect private property, which escalated the conflict. In rural areas that lacked a strong state presence, guerrilla and/or paramilitary groups were typically the main authorities who oversaw and established rules. This, along with unequal land distribution, weak land tenure, and poverty in rural Colombia, perpetuated the further appropriation of rural land, especially for small farmers and ethnic communities (Historical Memory Group, 2016). Particularly since the 1990s and 2000s, deregulated global markets and new laws have helped dismantle paramilitary groups. However, this has also created new opportunities for private agricultural and industrial projects to acquire land and natural resources in rural areas with weak tenure (Alfonso Sierra et al., 2011). These large industries have added complexities to tenure and environmental policy (Planeta Paz & Oxfam, 2017).

Today Colombia has the world's second-largest population of forcefully displaced people. According to the UNHCR, in 2018, there were estimated 8 million Colombians who had been forcefully displaced, all of whom experience increased tenure insecurity. In the case of Colombia, 98% of those forcefully displaced remain in Colombia (UNHCR, 2018). Several laws since the 2000s have recognized failures in Colombia's constitution and courts to prevent forced displacement and give victims of the armed conflict additional support mechanisms (Amnesty International, 2014; Ley 1448, 2011).

Fertile land coupled with weak land tenure and persistent underdevelopment in rural communities has furthermore enabled illicit crop production to thrive in Colombia, which has increased conflict, rural displacement, and environmental damage (Alvarez, 2007). According to the United Nations Office on Drugs and Crime, cocaine production levels are on the rise in Colombia, and production increased by 17% between 2016 and 2017 alone, thus demonstrating the continued need for tenure and resource security in rural areas (Oficina de Naciones Unidas Contra la Droga y el Delito, 2017).

Particularly since the Colombian Constitution of 1991, Indigenous and Afro-descendent ethnic communities have gained additional constitutional protection, recognizing ethnic groups' added vulnerabilities due to land insecurities. Around 3.4% of Colombia's population is identified as Indigenous, of whom 70% live in rural areas, whereas Afro-Colombians represent 10% of the national population (Amnesty International, 2014). The Colombian State allows ethnic populations to continue their traditional livelihoods and fully manage their land and resources in protected areas (Roldan Ortega, 2004). Such areas cover approximately 27% of the country and increase environmental conservation potential in addition to livelihood protection for ethnic groups (Amnesty International, 2014). Despite added protections, many ethnic communities still lack communal land titles, and weak security in rural areas increases their vulnerability to violence and exploitation.

In 2016, the Colombian government and the FARC-EP guerrilla group signed a Peace Agreement, outlining ambitious goals for achieving peace, reparations, and reconciliation. 'Land' was the first topic discussed at initial negotiations for the Peace Agreement, and 'Comprehensive Rural Reform', which aims for structural transformation and comprehensive development of rural Colombia, is the title of the first chapter in the final Peace Agreement (Guereña, 2017; Acuerdo Final, 2016). This chapter also prioritizes development programs in rural areas that have high poverty levels, weak institutions, high presence of illicit crops, and have been most affected by armed conflict (Acuerdo Final, 2016). However, the notion that the Peace Agreement has moved Colombia into a 'post-conflict' era is a misleading assumption which can leave already vulnerable populations and natural resources more vulnerable. While the Peace Agreement does present opportunities for a new path *toward* post-conflict (Latorre Restrepo, 2018), it also presents vulnerabilities that stem from newfound expectations, confrontation of past trauma, and shifts in power dynamics.

Despite the Agreement's commitment to comprehensive rural land reform, including mechanisms for restitution of land and formal titling, this complex issue is slow and difficult to change. In 2017, an estimated 79% of rural areas in Colombia were still without basic cadaster information (World Bank, 2017). Institutions in charge of land reform often lack sufficient resources to attend to the many cases coming forward, and information-sharing challenges between institutions further slow processes (Commission Étnica para la Paz y la Defensa de los Derechos Territoriales en Colombia, 2019). Countrywide there has been a dramatic increase in the number of assassinations and threats received by community leaders since the Peace Agreement was signed (Ojeda et al., 2015). Of the assassinated leaders, an estimated 70% have been smallholder farmers and/or from ethnic communities (El Tiempo, 2018). The continued violations of human rights throughout rural Colombia are intrinsically linked to the extent that land has been systemically devalued and natural resources destroyed. Land inequality in Colombia is growing, and Colombia continues to have the most unequal land distribution rates in Latin America (Guereña, 2017).

Climate Adaptation Amidst Tenure Confusion: The Case of Montes de María, Colombia

The complex land and conflict history in Colombia has direct implications for how the most socially vulnerable rural dwellers respond to climate stress. To understand the intersection of land tenure context and climate adaptation, we turn to the Montes de María region. The Montes de María are a series of low mountains in the southern part of the Bolivar Department on the Northeastern Caribbean coast of Colombia. This fertile region became a seat of intense fighting during the armed conflict, resulting in two cycles of violence and sweeping levels of internal displacement. First, from 1995 to 2005, the region experienced its highest levels of forced displacement and violence due to the army, paramilitary, and guerrilla groups battling in the same territory over control and ownership of land. Afterward, from 2005 to 2013, additional displacement and land tenure complexities were caused by the private sector, specifically extractive industries such as mining and agro-industrial projects, looking to take advantage of deregulated international markets and deescalated violence in rural areas. In the name of development and economic advancement, large areas of rural Colombia, including in Montes de Maria, were purchased by such industries. Land and the resources on it were often sold at cheap prices by 'secondary occupants' as the original owners had been displaced by conflict or were unable to gain a formal title to the land that they had previously held. Crops such as palm oil and teak tree plantations were planted throughout the territory and additional large swaths of land were acquired. Such projects have often been criticized for long-term environmental damage including loss of soil nutrients and depletion of water sources (Ojeda et al., 2015). In addition, there have been negative consequences on the health and livelihoods of rural farmers as their land and resources decreased and as they become wage laborers for large monocrops (Alfonso Sierra et al., 2011). By 2013, 43% or 74,000 hectares) of farmable land in the Montes de María were owned by investors from outside of the region and monocrop cultivation reached 100,000 hectares of land (Ojeda et al., 2015).

The majority of displaced farming families from the Montes de María returned to the area after active conflict calmed between 2010 and 2012, but this homecoming was met with problems of tenure insecurity. Rural properties that were formally owned tended to be large areas of land that were difficult for smallholder farmers to maintain through insecurity due to conflict (Planeta Paz & Oxfam, 2017). In addition, land ownership was typically inherited through family generations, often without formalized or registered tenure documents. Therefore, displaced families found themselves unable to demonstrate legal rights to their former land upon return if large companies or landowners had made claims to their property. Another common practice in the Montes de María was selling land in good faith through a signed note from the seller to the buyer; however, these documents do not hold up in a formal land dispute. Land restitution began with the signing of the Colombian Peace Agreement in late 2016. The Peace Agreements prioritized Montes de María for land restitution and, while more land titles have been formalized than in other regions, the overwhelming majority of claims remained in December of 2019. At the beginning of the drought, some families had submitted paperwork to formalize their tenure, and after the signing of the Peace Agreements, the majority of families began filing for their land to be restituted if they lost possession of the land, whether formal or informal, during the conflict.

In addition to land grabbing, displacement, and conflict, the Montes de María is already experiencing and anticipates more severe droughts, salinization of groundwater, shifts in precipitation, and increasing extreme heat that compound existing tenure insecurity (IPCC, 2018). The adaptation impacts of Colombia's complicated land tenure laws and conflict history came to the fore during a protracted drought in the Montes de María.[2] The region experienced the most severe drought on

[2] Evidence from farmers during the 2013–2016 El Niño drought is based on the qualitative fieldwork Castro conducted from 2016 through 2019 in Montes de María. Ethnographic data includes 130 in-depth interviews with farming families, local experts, and government officials, as well as copious field notes from ethnographic observation.

record due to an extended El Niño that lasted from 2013 through December 2016 (Cai et al., 2018). In the first year of drought, there were scarce rains and the majority of harvests failed. In 2015 and 2016, there was no rain at all and harvests were completely lost. The effects were particularly dire in this rural, developing region, which had no water or power infrastructure. Over the course of these three years of atypical drought and hot temperatures, farming families crossed all of the adaptation thresholds—adapting first in place and later through migration. When and how each family adapted depended on the security of their tenure, as well as their understanding of the land restitution process in the conflict recovery context.

In 2013, families first adapted to drought by waiting as long as possible to plant, but this proved problematic after seeds were lost through failed harvests when rain did not arrive for the entire 'rainy' season. Over the following four seasons, farmers planted on a more limited basis reducing the amount of land they planted and the variety in crops. Again, these harvests failed and farmers' seed stores ran out, food for consumption was limited, and assets liquidated. By 2015, as in situ adaptations had completely failed in the absence of precipitation to sustain rainfed agriculture, farmers began to adapt through migration. Adaptation through mobility depended on access to tenure and on understandings of tenure laws in this post-conflict context where the majority of farmers were in active land restitution processes. Families with legal title to their lands had higher tenure security: they were able to migrate temporarily and wait out the length of the drought, preserving their resources and reducing stress on their drought-affected land. Though doing so required selling livestock that also depended on rainwater, these families were able to leave without the perceived risk of losing their homes.

Families that had filed for land restitution, which would not begin until the Peace Agreement was signed, believed that their claims to land would be lost if they did not physically remain on that land. Even though the Office of Victims and the Office of Land in Colombia said that this was not the case, families were unconvinced that their land tenure claims would be honored if they displaced due to drought. Farmers also expressed fears of land grabs by corporations, other farmers, or illicit crop producers. As a result, these families depleted their financial and social capital while remaining on dried out land for the duration of the El Niño drought. This effort to remain in place came at a high cost—requiring families to liquidate their limited assets, remove children from school because they could not afford transportation or send them walking without sufficient water or food in the heat, and take out loans from banks to buy seeds and attempt to plant over and over each season leaving them indebted after the drought ended. This was the most common case for subsistence farming families in the Montes de María region, who made due in place by exhausting their resources in order to stay on their land parcel and continue their unresolved land tenure claims. Furthermore, unresolved tenure claims prevented families from making investments in their properties that would have fostered their resilience to climate hazards, such as digging water reservoirs, installing mechanized pumps to deliver water to fields and animals, and installing permanent dwellings that would provide more protection from high temperatures. Tenure confusion ultimately generated a highly vulnerable, trapped population in the Montes de María.

The misalignment of government and community priorities is evident in the way land restitution is administered—through titling plots throughout the drought-prone region and allotting livestock, specifically resource-intensive cows, to farmers in an already-water-stressed region. In Colombia more generally, more than double the amount of land deemed suitable for livestock is already being used (Guereña, 2017). In the first two years of land restitution and implementation of the Peace Agreement, farming families described their desire to move to land less prone to drought and closer to freshwater access, but also noted their inability to do so due to their land restitution claims tied to the Montes de María. Many farmers opted to begin pig raising operations as opposed to cows after the drought because pigs require less water and less land than cows, then received cattle as part of their land restitution package in 2018 and 2019 and began wrestling with how they would secure water and pasture for cattle as local rainfall remains low. Even homes provided by the land restitution process came without water storage as part of their design. Across the board, lessons for climate adaptation that farming families learned the hard way during the El Niño drought are actively undermined by national land restitution and conflict recovery policies that fail to consider the climate realities of the Montes de María.

Looking Forward: Considering Land Tenure and Climate Adaptation Together

A close look at the Colombian case highlights the need for more effective and long-term global sustainable development policy that considers land resources, land tenure, and climate change jointly in order to be effective in today's climate change reality. Land tenure policy could be improved in two ways—by flexibly embracing and incorporating environmental rights and citizen participation. Improving land tenure security is key to enabling responsive, local-level autonomous climate adaption.

We argue that socio-environmental vulnerability can only be mitigated through regarding the natural environment as deserving of rights and protection. Regarding 'land' as a living being which also merits rights could bolster the protection and prosperity of natural ecosystems and all organisms that depend on them (Acosta, 2019). In the face of climate change, humans need environmental rights to foster sustainable development. However, the interconnection of human rights with environmental rights is largely absent in the Universal Declaration of Human Rights (UN General Assembly, 1948). Policies across fields should incorporate environmental rights not just because of international pressure, but also because considering a cost-benefit analysis of development without environmental protection demonstrates a high potential of harm to natural and human systems. The Sustainable Development Goals are an example of one such framework that incorporates environmental rights and protections as part of broader development goals. The Colombian case demonstrates the need for flexible policy that can be responsive to a variety of locally occurring adaptation mechanisms in different contexts and climate stressors (Unruh & Abdul-Jalil, 2012).

The other side of flexibly incorporating environmental rights into all realms of development policy is the need for flexible citizen engagement requirements in land policy processes. Realities of land tenure, demographics, and natural resources differ greatly depending on location and, furthermore, as climate change manifests in different contexts, the relationships between these variables are hyperlocal and are in flux (Quan & Dyer, 2008). In order to account for such differences, it is paramount that policy, including for land tenure and climate change, contains citizen engagement[3] and participation requirements. If higher-level policy can incorporate such priorities, the pressing needs of the most vulnerable in addition to root causes of their grievances will be more effectively addressed.

Within international development organizations, there has been a movement to include citizen engagement mechanisms for project beneficiaries. The World Bank, for example, which plays a role in land tenure projects in Colombia and around the world, outlines requirements in its *Strategic Framework for Mainstreaming Citizen Engagement in World Bank Group Operations*. Here, the World Bank mandates citizen engagement in 100% of its projects with 'clearly identified beneficiaries' and aims to achieve this strategy by "empowering citizens to participate in the development process and integrating citizen voice in development programs as key accelerators to achieving results" (World Bank Group, 2014). Such requirements are significant in themselves for increasing the likelihood for citizen engagement; however, they must also be flexible and adaptive to local contexts (Fox, 2014). Community consultations should be coupled with 'hard' accountability mechanisms such as sanctions and answerability requirements in order to increase the likelihood of effective participation and feedback (Fox, 2007). Furthermore, instead of focusing reporting upward to donors or government agencies, sharing mechanisms which give results outwardly to citizens also help perpetuate project accountability. Undoubtedly, international organizations and their policies have a large role to play in citizen engagement. Land tenure projects that include effective citizen engagement mechanisms will be more likely to account for climate-based needs and foster adaptive livelihoods in rural areas.

---

[3] *Citizen engagement* is 'the two-way interaction between citizens and governments or the private sector within the scope of interventions that gives citizens a stake in decision-making with the objective of improving the intermediate and final development outcomes of the intervention' (World Bank Group, 2014).

In sum, concrete yet flexible land and development policy attuned to environmental realities could protect the rights of rural dwellers in the developing world, making populations less socio-environmentally vulnerable and thereby facilitating successful local autonomous adaptation to climate stress. Less vulnerable people are better able to adapt to their environment and utilize land resources in the ways that works for them. This adaptability requires rural, resource-dependent populations to be nimble in the face of shifting environmental conditions. Current development and land policies are rigid, trapping populations in adaptations that may not suffice in the era of climate change.

References

Acosta, A. (2019). Amnesty International. https://nextstrategy.amnesty.org/2019/06/20/sin-derechos-de-la-naturaleza/

Acuerdo Final para la Terminación del Conflicto y la Construcción de una Paz Estable Y Duradera. (2016). Colombian Government—FARC EP, November.

Adger, W., Hug, S., Brown, K., Conway, D., & Hulme, M. (2003). Adaptation to climate change in the developing world. *Progress in Development Studies, 3*(3), 179–195.

Adger, W. N., Lorenzoni, I., & O'Brien, K. L. (2009). Adaptation now. In W. Neil Adger, I. Lorenzoni, & K. L. O'Brien (Eds.), *Adapting to climate change: Thresholds, values, governance* (pp. 1–22). Cambridge University Press.

Ahmed, K. F., Wang, G., You, L., & Yu, M. (2016). Potential impact of climate and socioeconomic changes on future agricultural land use in West Africa. *Earth System Dynamics, 7*(1), 151–165. https://doi.org/10.5194/esd-7-151-2016

Albert, S., Leon, J. X., Grinham, A. R., Church, J. A., Gibbes, B. R., & Woodroffe, C. D. (2016). Interactions between sea-level rise and wave exposure on reef island dynamics in the Solomon Islands. *Environmental Research Letters, 11*(5), 054011.

Alfonso Sierra, T., Grueso Castelblanco, L., Prada Rivas, M., & Salinas Abdala, Y. (2011). Derechos enterrados: Comunidades étnicas y campesinas en Colombia, nueve casos de estudio. Bogota; Universidad de los Andes, Facultad de Derecho, Centro de Investigaciones Sociojurídicas. https://repositorio.uniandes.edu.co/bitstream/handle/1992/8204/derechos-enterrados-comunidades-etnicas.pdf

Althor, G., Watson, J. E., & Fuller, R. A. (2016). Global mismatch between greenhouse gas emissions and the burden of climate change. *Scientific Reports, 6*, 20281.

Alvarez, M. (2007). Environmental damage from illicit crops in Colombia. In *Extreme conflict and tropical forests* (pp. 133–147).

Amnesty International. (2014). A land title is not enough: Assuring sustainable land restitution in Colombia. London: Amnesty International Ltd. http://www.oidhaco.org/uploaded/content/article/1975836477.pdf

Andrew D., King Luke J., Harrington. (2018). The Inequality of Climate Change From 1.5 to 2°C of Global Warming. *Geophysical Research Letters, 45*(10), 5030–5033. https://doi.org/10.1029/2018GL078430

Bardsley, D. K., & Hugo, G. J. (2010). Migration and climate change: Examining thresholds of change to guide effective adaptation decision-making. *Population and Environment, 32*(2–3), 238–262.

Black, R., Arnell, N. W., Adger, W. N., Thomas, D., & Geddes, A. (2013). Migration, immobility and displacement outcomes following extreme events. *Environmental Science & Policy, 27*, S32–S43.

Brown, H. C. P., & Sonwa, D. J. (2015). Rural local institutions and climate change adaptation in forest communities in Cameroon. *Ecology and Society, 20*(2), 6.

Cai, W., Wang, G., Dewitte, B., Wu, L., Santoso, A., Takahashi, K., Yang, Y., Carréric, A., & McPhaden, M. J. (2018). Increased variability of eastern Pacific El Niño under greenhouse warming. *Nature, 564*(7735), 201.

Castro, B., (2019). The shifting limits of drought adaptation in rural Colombia. In *Current Directions in Water Scarcity Research* (Vol. 2, pp. 77–86). Elsevier.

Commission Étnica para la Paz y la Defensa de los Derechos Territoriales en Colombia. (2019). 1er informe de cumplimiento del capítulo étnico. Medellin: Taller de Artes Grafica Periferia. https://www.onic.org.co/images/pdf/1er_informe_cumplimiento_cap_etnico_paz.pdf

Del Pilar Lopez-Uribe, M., & Sanchez Torres, F. (2018). On the agrarian origins of civil conflict in Colombia. Latin America and Caribbean Center.

Dumenu, W. K., & Obeng, E. A. (2016). Climate change and rural communities in Ghana: Social vulnerability, impacts, adaptations and policy implications. *Environmental Science & Policy, 55*, 208–217.

El Tiempo. (2018). *En tres meses, 120 líderes sociales han sido asesinados en Colombia*. El Tiempo. https://www.eltiempo.com/colombia/otras-ciudades/elmapa-de-los-lideres-sociales-asesinados-en-colombia-184408

Fankhauser, S., Smith, J., & Tol, R. (1999). Weathering climate change: Some simple rules to guide adaptation decisions. *Ecological Economics, 30*(1), 67–78.

Fox, J. (2007). The uncertain relationship between transparency and accountability. *Development in Practice, 17*(4–5). https://escholarship.org/uc/item/8c25c3z4

Fox, J. (2014). Social accountability: What does the evidence really say? *World Development, 72*, 346–361.

Guereña, A. (2017). A snapshot of inequality: What the latest agricultural census reveals about land distribution in Colombia. Oxfam. https://www-cdn.oxfam.org/s3fs-public/file_attachments/colombia_-_snapshot_of_inequality.pdf

Historical Memory Group. (2016). Basta Ya! Colombia: Memories of war and dignity. Bogota: National Center for Historical Memory.

Hoegh-Guldberg, O., Jacob, D., Taylor, M., Bindi, M., Brown, S., Camilloni, I., Diedhiou, A., Djalante, R., Ebi, K. L., Engelbrecht, F., Guiot, J., Hijioka, Y., Mehrotra, S., Payne, A., Seneviratne, S. I., Thomas, A., Warren, R., & Zhou, G.. (2018). Impacts of 1.5°C global warming on natural and human systems. In V. Masson-Delmotte, P. Zhai, H.-O. Pörtner, D. Roberts, J. Skea, P. R. Shukla, A. Pirani, W. Moufouma-Okia, C. Péan, R. Pidcock, S. Connors, J. B. R. Matthews, Y. Chen, X. Zhou, M. I. Gomis, E. Lonnoy, T. Maycock, M. Tignor, & T. Waterfield (Eds.), *Global Warming of 1.5°C. An IPCC Special Report on the impacts of global warming of 1.5°C above pre-industrial levels and related global greenhouse gas emission pathways, in the context of strengthening the global response to the threat of climate change, sustainable development, and efforts to eradicate poverty.*

IPCC. (2018). Global warming of 1.5°C. In V. Masson-Delmotte, P. Zhai, H. O. Pörtner, D. Roberts, J. Skea, P. R. Shukla, A. Pirani, W. Moufouma-Okia, C. Péan, R. Pidcock, S. Connors, J. B. R. Matthews, Y. Chen, X. Zhou, M. I. Gomis, E. Lonnoy, T. Maycock, M. Tignor, & T. Waterfield (Eds.), *An IPCC Special Report on the impacts of global warming of 1.5°C above pre-industrial levels and related global greenhouse gas emission pathways, in the context of strengthening the global response to the threat of climate change, sustainable development, and efforts to eradicate poverty.*

Latorre Restrepo, S. (2018). Conflicto armado y transición hacia posconflicto: una aproximación desde el Caribe. Barranquilla: Editorial Universidad del Norte.

Ley 1448. (2011). El Congreso de la República. https://www.unidadvictimas.gov.co/sites/default/files/documentosbiblioteca/ley-1448-de-2011.pdf

Masuda, Y. J., Castro, B., Aggraeni, I., Wolff, N. H., Ebi, K., Garg, T., & Spector, J. (2019). How are healthy, working populations affected by increasing temperatures in the tropics? Implications for climate change adaptation policies. *Global environmental change, 56*, 29–40.

Mertz, O., Mbow, C., Reenberg, A., & Diouf, A. (2009). Farmers' perceptions of climate change and agricultural adaptation strategies in rural Sahel. *Environmental Management, 43*(5), 804–816.

Meze-Hausken, E. (2008). On the (im-) possibilities of defining human climate thresholds. *Climatic Change, 89*(3–4), 299–324.

Mortimore, M. J., & Adams, W. M. (2001). Farmer adaptation, change and 'crisis' in the Sahel. *Global Environmental Change, 11*(1), 49–57.

Morton, J. F. (2007). The impact of climate change on smallholder and subsistence agriculture. *Proceedings of the National Academy of Sciences, 104*(50), 19680–19685.

Neumann, B., Vafeidis, A. T., Zimmermann, J., & Nicholls, R. J. (2015). Future coastal population growth and exposure to sea-level rise and coastal flooding-a global assessment. *PLOS One, 10*(3), e0118571.

O'Brien, K. L., & Leichenko, R. M. (2000). Double exposure: Assessing the impacts of climate change within the context of economic globalization. *Global Environmental Change, 10*(3), 221–232.

Oficina de Naciones Unidas Contra la Droga y el Delito. (2017). Informe de monitoreo de territorios afectados por cultivos ilícitos 2017. Bogota: SIMCI—UNODC. https://www.unodc.org/documents/crop-monitoring/Colombia/Colombia_Monitoreo_territorios_afectados_cultivos_ilicitos_2017_Resumen.pdf

Ojeda, D., Petzl, J., Quiroga, C., Rodríguez, A. C., & Rojas, J. G. (2015). Paisajes del despojo cotidiano: acaparamiento de tierra y agua en Montes de María, Colombia. *Revista de Estudios Sociales*, (54), 107–119. https://doi.org/10.7440/res54.2015.08

Paprocki, K. (2018). Threatening dystopias: Development and adaptation regimes in Bangladesh. *Annals of the American Association of Geographers, 108*(4), 955–973.

Planeta Paz & Oxfam. (2017). Agenda Común Para la Paz desde los Territorios. Bogota: Espacio Creativo Impresores SAS. https://drive.google.com/file/d/1OnsyreNF3V56Cby6YhWMzXFC0VFShJiE/view?usp=sharing

Quan, J., & Dyer, N. (2008). Climate change and land tenure: The implications of climate change for land tenure and land policy. *Land Tenure Working Paper 2*. University of Greenwich. http://www.fao.org/3/a-aj332e.pdf

Reij, C., & Waters-Bayer, A. (2014). *Farmer innovation in Africa: A source of inspiration for agricultural development*. Routledge.

Roldan Ortega, R. (2004). Models for recognizing indigenous land rights in Latin America. The International Bank for Reconstruction and Development/The World Bank.

Schewel, K. (2019). Understanding immobility: Moving beyond the mobility bias in migration studies. *International Migration Review*. https://doi.org/10.1177/0197918319831952

Smit, B., Burton, I., Klein, R., & Wandel, J. (2000). An anatomy of adaptation to climate change and variability. *Climatic Change, 45*(1), 223–251.

Stark, O., & Bloom, D. E. (1985). The new economics of labor migration. *The American Economic Review, 75*(2), 173–178.

Sunderlin, W. D., de Sassi, C., Sills, E. O., Duchelle, A. E., Larson, A. M., Resosudarmo, I. A. P., et al. (2018). Creating an appropriate tenure foundation for REDD+: The record to date and prospects for the future. *World Development, 106*, 376–392. https://doi.org/10.1016/j.worlddev.2018.01.010

Thomas, D. S., & Twyman, C. (2005). Equity and justice in climate change adaptation amongst natural-resource-dependent societies. *Global Environmental Change, 15*(2), 115–124.

UN General Assembly. (1948). *The Universal Declaration of Human Rights*.

UNHCR. (2018). Global trends: Forced displacement in 2018.

United Nations. (2018). *The 2030 Agenda and the Sustainable Development Goals: An opportunity for Latin America and the Caribbean* (LC/G. 2681-P/Rev. 3), Santiago.

Unruh, J., & Abdul-Jalil, M. A. (2012). Land rights in Darfur: Institutional flexibility, policy and adaptation to environmental change. *Natural Resources Forum, 36*, 274–284.

World Bank. (2017). Análisis Ambiental y Social y Marco de Gestión Ambiental y Social: Análisis Ambiental y Social y Marco de Gestión Ambiental y Social. Colombia: s.n. http://documents.worldbank.org/curated/en/456891542191758619/pdf/An-lisis-y-Marco-de-Gesti-n-Ambiental-y-Social-Recursos-Culturales-Fisicos.pdf

World Bank Group. (2014). Strategic framework for mainstreaming citizen engagement in World Bank Group operations. http://documents.worldbank.org/curated/en/266371468124780089/071652160_201411345005606/additional/929570WP0Box380ategicFrameworkforCE.pdf

Yuta, J., Masuda Brianna, Castro Ike, Aggraeni Nicholas H., Wolff Kristie, Ebi Teevrat, Garg Edward T., Game Jennifer, Krenz June, Spector. (2019). How are healthy working populations affected by increasing temperatures in the tropics? Implications for climate change adaptation policies. *Global Environmental Change, 55*, 1–14. https://doi.org/10.1016/j.gloenvcha.2019.03.005

10

Interaction of Conditional Incentives for Ecosystem Conservation with Tenure Security: *Multiple Roles for Tenure Interventions*

Iliana Monterroso and Erin Sills

The Role of Tenure in PES Programs

Property rights are central to the concept of conditional incentives, such as used in agri-environmental programs that pay farmers to conserve natural resources. In the Global South, this policy instrument has become known as PES, or payment for ecosystem services, and has inspired REDD+, or Reducing Emissions from Deforestation and Forest Degradation. The potential for REDD+ to make significant contributions to both climate change mitigation and conservation of tropical forests has focused attention on securing forest land tenure to enable conditionality.

I. Monterroso (✉)
Center for International Forestry Research (CIFOR), Bogor, Indonesia
e-mail: i.monterroso@cgiar.org

E. Sills
North Carolina State University, Raleigh, NC, USA

M. B. Holland et al. (eds.), *Land Tenure Security and Sustainable Development*,
https://doi.org/10.1007/978-3-030-81881-4_10

We examine the evidence on how tenure security interacts with participation in PES by individual land stewards, showing that the relationship is multidimensional and bi-directional. We then consider the relationship between tenure systems and implementation of REDD+, which depends on the type of conditionality and the type of tenure challenge faced. Secure tenure is widely understood to be a necessary enabling condition for the implementation of REDD+, but we show that tenure security can also be an integral part of conditional incentives. We illustrate that with the tenure interventions supported by the Dedicated Grant Mechanism (DGM) for Indigenous Peoples and Local Communities (IPLC) in Peru and Indonesia. The DGM sought to clarify tenure rights and strengthen tenure security over forest resources for both Indigenous and other rural populations, but applied different concepts of conditionality to different groups. In both Indonesia and Peru, conditionality was embedded in the communal land titles that recognize customary tenure rights, while the social forestry program in Indonesia was implemented as both a prerequisite and a reward for participation in REDD+ in state forests.

Direct conditional incentives—often monetary payments—to forest stewards have been promoted as a way to sustain the provision of critical ecosystem services in forest landscapes (Duchelle et al., 2018; Robinson et al., 2018). The idea is that paying people directly for the provision of ecosystem services is the most assured and efficient way to secure sustained conservation of these ecosystems over time (Ferraro & Kiss, 2002; Bruce et al., 2010). In theory, recognizing the value of public goods and services, and creating economic incentives for their protection should safeguard them (Engel et al., 2008; Wunder et al., 2008). In practice, the additionality of PES programs may be undercut by factors such as participation by landowners who would have conserved ecosystems even without payment. The limited evidence available on effectiveness comes primarily from evaluations of the PES programs in Costa Rica and Mexico, which find either no impact or very small positive impacts of participation on forest cover (Samii et al., 2014; Alix-Garcia & Wolff, 2014).

Despite this limited evidence, the core principles behind PES were rapidly adopted in what was expected to become the largest international conservation scheme for tropical forests: REDD+. According to Sunderlin et al. (2018), REDD+ was "to create conditional incentives based on performance to prevent forest conversion (REDD) and for enhancing forest carbon stocks (the +)." Alternative strategies for implementing REDD+ have been tested in hundreds of REDD+ pilot projects across the Global South (Simonet et al., 2018). Fewer than half of the projects offered direct, conditional payments for forest conservation to land stewards (Wunder et al., 2020). Drawing on data from a sub-sample of these pilot projects, Wunder et al. (2020) report that conditional incentives were widely considered the most effective tool for promoting forest conservation, by both implementing organizations and land stewards. However, many implementing organizations did not expect to use conditional incentives, both because of the lack of secure long-term financial flows for climate change mitigation and because the pervasive insecure land tenure impedes effective contracting with land stewards.

Research has confirmed the importance of clear and secure land and forest tenure for the effectiveness of incentive-based instruments at improving natural resource management (Agrawal et al., 2014; Galik & Jagger, 2015; Larson et al., 2013; Robinson et al., 2014; Sunderlin et al., 2018). Lack of exclusion rights in particular can undermine the effectiveness of PES contracts (Clements et al., 2010), although that also depends on the structure of the PES program (e.g. see Jones, MacDonald, et al., 2020). For example, Rosales (2003) describes PES programs that formalized and recognized customary tenure in the Philippines. In fact, tenure security may effectively be the incentive offered for ecosystem conservation (e.g. in Ecuador as discussed by Buntaine et al., 2015 and Holland et al., 2017). Thus, the security of land tenure can both influence who participates and be influenced by participation in PES (Börner et al., 2017, 2011; Holland et al., 2014; Smith et al., 2017; Swallow & Meinzen-Dick, 2009). Some PES programs require that participants have land titles (Bremer et al., 2014; Jones, Etchart, et al., 2020). Even in those cases, participation in the program can lend legitimacy and increase the security of land tenure (e.g. in Costa Rica as discussed by Arriagada et al., 2009; Miranda et al., 2003). Jones, Etchart, et al. (2020) find that participation in a PES program reduced conflict over land under *de facto* communal tenure, but not in communities with *de facto* private land tenure in Ecuador. In China, Liu et al. (2018) also found that the effects of participation on tenure depend on the initial tenure conditions as well as the institutional details of the program.

The Wide Spectrum of Conditionality and Tenure Security

Given the multiple and critical roles of tenure in conditional incentives, interest in using PES for REDD+ has helped motivate interventions to address long-standing tenure insecurities, for example, through the "Terra Legal" program in Brazil (Duchelle et al., 2017) and the "One Map Policy" in Indonesia (Astuti & McGregor, 2015; Mulyani & Jepson, 2016; Resosudarmo et al., 2014; Sunderlin et al., 2018). The increased attention to tenure raises questions about the optimal sequencing, integration, and design of tenure interventions and conditional incentives. This depends in part on the specific forms of conditionality and tenure insecurity.

Conditionality is critical for securing service provision in PES systems (Engel et al., 2008). In PES designed to mimic market exchange of ecosystem services, conditionality means that rewards or benefits received by the ecosystem service (ES) provider are conditional on compliance or performance measures agreed in contracts between parties (van Noordwijk & Leimona, 2010). However, the degree and form of conditionality incorporated into PES programs vary widely (Hejnowicz et al., 2014). Conditionality can take the form of either incentives or rewards, delivered either by market-based instruments or by public programs (see Box 10.1). The conditions can be long term such as sustained provision of environmental services, medium term such as engaging or foregoing a particular resource use, or short term such as participation (Engel et al., 2008). Regardless, conditionality requires establishing systems for monitoring, enforcement, and sanctions (Newton et al., 2012). State actors play particularly important roles in enforcing conditionality for ES that are public goods (e.g. biodiversity, carbon sequestration) and in cases with strong incentives for free-riding. Figueroa et al. (2016) argue that the observed variation in conditionality reflects variation in socio-economic

Box 10.1 Identifying Different Levels of Conditionality

Van Noordwijk and Leimona (2010, pp. 6–9) identify different levels of conditionality:

- **Level 1** is based on actual service delivery and direct marketability of a commoditized environmental service, for example, carbon credits. This refers to direct market transactions between service providers and buyers.
- **Levels 2 and 3** operate in the context of compensation of opportunities foregone and are based on the achievement of an objectively measurable condition of the agricultural or forest landscape. This refers to financial compensation of opportunity costs (by private or public actors).
- **Level 4** emerges in the context of co-investments in the landscape and may include negotiated tenure, investment in public services, or land use planning conditional of ES maintenance. This level of conditionality is based on trust in local communities to enhance ES provision under flexible contracts in the presence of monitoring and sanctions. Benefits might not be directly linked with ES provision and financial opportunity costs might not be fully paid.

and political conditions, such as forest and land management practices, livelihood strategies of forest dwellers, social differentiation, migration, and the structure and processes of governance.

Based on their analysis of PES programs across the Global South, van Noordwijk and Leimona (2010) conclude that Level 1 conditionality is often not appropriate because enhancement of ES cannot be disentangled from development needs, especially in the context of unclear, overlapping, and contested rights to natural resource rights. They call for review of existing legal frameworks to identify and establish the appropriate level of conditionality. Likewise, Newton et al. (2012) suggest that Bolsa Floresta in Brazil would be more effective if its payment structure were adjusted to account for different opportunity costs and livelihood strategies. In Indonesia, Kerr et al. (2014) recommend that the conditionality framework for community forests, or *Hutan Kemasyarakatan*, should be based not only on the delivery of environmental service, but also on the maintenance of the ecosystem in a desirable state and development of institutional arrangements that further enhance ES service provision. The existing conditionality framework calls for eviction of stakeholders who do not abide by the contract terms, which is politically unrealistic.

Conditionality is fundamentally based on property rights, which determine who is eligible to receive an incentive, and therefore who benefits, who is excluded, and who is responsible and held accountable for meeting contract obligations. Absent or weak property rights can prevent resource stewards from participating in PES schemes and REDD+ initiatives (Blackman et al., 2017b; Wunder, 2013; Wunder et al., 2008). Recognition and distribution of property rights is not a straightforward process. First, even in situations where statutory rights are fully transferred from public ownership to individuals or collectives (e.g. land titles of Indigenous territories as described in Chap. 4), the government can place restrictions or conditions on the bundle of rights. For instance, pro-forest conditions linked to titling programs may include the obligation to forgo forest-clearing activities or maintaining a portion of land in forests (Bruce et al., 2010). Second, only a partial bundle of rights may be recognized among groups of resource users (e.g. via co-management agreements). Third, the State may grant rights to different resources in the same territory via concessions or licenses, for example, an extractive concession granted within the boundaries of an established protected area and/or Indigenous territory (Monterroso et al., 2019). To address this range of possibilities, Sunderlin et al. (2018) and Sunderlin, Larson, et al.

Box 10.2 Early Tenure Actions in the Context of Incentive-Based Initiatives

Objectives of tenure interventions for conditional incentives

1. Clarify rights. Tenure arrangements determine who benefits; therefore, any initiative needs to define clearly who are the right-holders for rewards and incentives.
2. Establish responsibilities/accountabilities. Tenure arrangements determine responsibilities and accountabilities. This includes clarifying inter-sectoral and inter-ministerial tenure contestation at all scales.
3. Avoid resource rush.
4. Minimize negative effects of actions on local livelihoods and rights (resource use restrictions).
5. Strengthen the ability to exclude outsiders (provision of enforceable rights of exclusion).

Source: Buntaine et al. (2015), Sunderlin et al. (2018) and Sunderlin, Larson, et al., 2014)

(2014) argue that tenure must be considered holistically and from the beginning, for example, in readiness strategies that lay the groundwork for REDD+ (Box 10.2).

Clear and uncontested property rights allow ES suppliers to meet obligations and ES buyers to enforce contract commitments (Bruce et al., 2010; Naughton-Treves & Wendland, 2014; Resosudarmo et al., 2014; Robinson et al., 2014). Rights provide the authority to make land-use decisions and ensure protection against external claims. Both are often necessary to meet the conditions established for an incentive such as PES. Enforcement of existing rights requires sound monitoring and sanctioning rules as well as harmonized and clear implementation procedures in place in cases of infractions (Bruce et al., 2010; Naughton-Treves & Wendland, 2014; Robinson et al., 2014). Thus, broader legal and socio-political support including inter-sectorial coordination and collaboration are required to ensure the robustness of rights.

Tenure interventions should be tailored to the particular tenure challenge, that is, whether rights are unclear, insecure, or in conflict (Table 10.1). Addressing these tenure challenges is a highly contested and political process (Naughton-Treves & Wendland, 2014). While land titling is widely considered to provide the greatest tenure security, some interventions have recognized different sub-sets of the full bundle of rights (access, management, exclusion) to different sub-sets of the resources or services associated with land (e.g. wood, non-wood forest products, carbon rights, and water provision) (Bruce et al., 2010; Naughton-Treves & Wendland, 2014).

Tenure Interventions in the Context of the Dedicated Grant Mechanism

The DGM was established in 2010 to support the full and effective participation of IPLC in REDD+. Critics of REDD+ have long pointed out that conditional incentives like PES are difficult—if not impossible—to implement where resource tenure is unclear and highly contested, as is broadly the case in countries in the Global South (Naughton-Treves & Wendland, 2014; Sunderlin, Ekaputri, et al., 2014; Sunderlin, Larson,

Table 10.1 Characterization of tenure interventions in the context of conditional incentives

Tenure challenge	Type of intervention/goals	Example of interventions
Rights unclear	Clarification of tenure rights or right-based approaches, as suggested by Agrawal et al. (2014), vary depending on whether they promote: 1. Creation of new rights 2. Modifying the type of right 3. Reallocating resource rights to different right holder	• Land titling • Demarcation and mapping of village/land/forests boundaries • Documentation and registration of rights in public registries (e.g. cadaster) • Review of existing overlapping rights (and claims) through regularization/formalization • Reforms in legislations to recognize/reallocate/clarify/modify rights • Establishment of protected areas (or setting aside protection areas) to modify land uses • Social forestry schemes that recognize community forest management rights
Rights insecure	Interventions to strengthen and enforce the robustness and guarantee of rights	• Enforcement of exclusion rights through monitoring and sanctioning rules • Legal and socio-political support of resource rights • Harmonize or clarify procedures and rules • Review implementation processes (identify overlapping mandates)—inter-sectorial coordination and collaboration
Rights in conflict	Mechanisms that enforce and protect the exercise of rights in conflict situations	• Conflict management and conflict resolution mechanisms • Harmonization or clarification of procedures and rules • Grievance mechanisms (including compensation) • Review overlapping mandates across government institutions • Enforcement of monitoring and sanctioning rules

Sources: Agrawal et al. (2014), Blackman et al. (2017b), Blackman and Veit (2018), Bruce et al. (2010), Buntaine et al. (2015), Holland et al. (2014), Larson et al. (2013), Naughton-Treves and Wendland (2014), Robinson et al. (2017) and Smith et al. (2017)

et al., 2014). The tropical forest regions of critical importance for REDD+ have complex and overlapping tenure regimes, where often what is legally or formally declared in terms of tenure does not match with the reality of tenure as defined or recognized among communities. IPLC are estimated to hold tenure rights to as much as 65% of forest in developing countries, but only 18% of this land is formally recognized either as owned or designated for their use (RRI, 2015). The vast majority of forest lands are officially owned by governments. These overlapping tenure systems affect not only communities and governments, but also private sector investors and owners (Sunderlin et al., 2014).

In addition to being a barrier to conditional payment schemes, overlapping and insecure tenure has been identified in national REDD+ readiness processes as a key driver of deforestation and ecosystem degradation (Sunderlin and Larson, et al., 2014). Effectively, tenure insecurity makes it harder to address the business-as-usual drivers of deforestation. Additionally, given the substantial amount of funding expected for REDD+ and for carbon credits more generally, there have been concerns that lack of clear tenure would encourage a type of resource rush or "land grab" to make carbon deals and capture REDD+ funding, leading to dispossession of traditional and customary landholders. Limiting participation to those with formal land titles could also bake in historical inequalities and exclude IPLC (Broegaard et al., 2017; Chomba et al., 2016; Johnson et al., 2018; Samndong & Vatn, 2018).

Concerns over these risks mobilized a movement for "no rights no REDD+" (Howell, 2014). Promoted on the ground by Indigenous and traditional peoples (Myers et al., 2017, 2018), this movement called for the adoption of specific measures that favored institutional changes through tenure clarification (Duchelle et al., 2018) and other types of rights-based approaches (Agrawal et al., 2014). As a result of the attention to tenure in both the scientific literature and popular movements, substantial funding for REDD+ readiness has been allocated to tenure interventions. This includes the DGM, which is supporting national programs to clarify communal and customary land tenure in order to establish the conditions for collective conditional incentives for IPLC.

Among the 13 countries targeted by the DGM, Peru and Indonesia have been subject to the most research, which we review for insight into the multiple roles of tenure interventions in REDD+ (Blackman et al.,

2017a; Blackman & Veit, 2018; Duchelle et al., 2017; Resosudarmo et al., 2014; Sunderlin et al., 2018). Specifically, we characterize the tenure interventions by identifying the goals of the proposed reforms, the tenure regimes targeted, the content of the proposed reforms, and the stakeholders involved. In this context, we identify two distinct roles for tenure, corresponding to two versions of conditionality.

Over the past decade, the DGM has supported a wide diversity of actions to facilitate incentive-based REDD+, notably including many tenure interventions as summarized in Table 10.2 for Peru and Indonesia. The DGM explicitly uses REDD+ to leverage interventions to secure land rights, thus helping to avoid further forest conversion and conflicts over incentives (DGM, 2019, p. 25). In both Indonesia and Peru, this includes formal recognition of the customary rights of Indigenous People. Consistent with the general belief and limited scientific evidence that Indigenous People conserve forests that they own (Nepstad et al., 2006; Nelson & Chomitz, 2011), both of these interventions embedded conditionality into the tenure instruments themselves, conditioning tenure on forest stewardship. Thus, recognition of tenure both enabled and functioned as the conditional incentive for forest conservation.

In the Peruvian Amazon, where large forest areas are held by Indigenous Peoples, titling has been promoted as a critical enabling condition for national REDD+ initiatives (Blackman & Veit, 2018; Evans et al., 2014; Robinson et al., 2017). While titling of indigenous lands started in the late 1970s, it stalled for decades due to lack of political support, changes in the institutional framework, and cumbersome procedures. In 2014, during COP 21, international supporters called for action to overcome challenges including lack of financial support to complete the regularization of communities in target areas (Monterroso et al., 2017). Since 2015, international funding has flowed into multiple environmental projects that also support the recognition, demarcation, and titling of the communal land holdings of native communities (Monterroso & Larson, 2018). One of these projects was the Saweto DGM, which was allocated USD5.5 million under the Forest Investment Program (administered by the World Bank) and supported the recognition of 310 native communities and the demarcation and titling of almost 1 million hectares in the Amazon (Sunderlin et al., 2018).

Table 10.2 Tenure interventions in the context of DGM schemes in Peru and Indonesia

Country	Tenure intervention and proposed actions	Changes in the bundle of rights	Level of conditionality
Peru	DGM Saweto focused on clarifying and securing Indigenous communities' tenure rights through: • Legal recognition of native communities • Demarcation of communal villages and forests and documentation of existing rights • Granting of collective property titles in agricultural lands • Granting of usufruct contracts in forest lands • Promotion of community forest management	• Communal land titles recognize decision-making rights over agricultural land • Usufruct rights to community forests are granted in perpetuity, but management of those areas must comply with other regulatory procedures (e.g. submission of management plans, logging permits) • State retains alienation rights and rights to subsoil (minerals and oil)	**Level 4. Communal Land Titling**. Tenure requires maintenance of ES

(*continued*)

Table 10.2 (continued)

Country	Tenure intervention and proposed actions	Changes in the bundle of rights	Level of conditionality
Indonesia	Formalization of customary rights vary depending whether they are implemented in: Private Forests • Collective rights to lands and forests within ancestral territories of customary peoples (*masyarakat hukum adat*). State Forests • Community forest licenses (IUP-HKM[a]) granted local communities or groups adjacent to state forests classified as both production and protection forests. • HTR[a] permits granted to communities, associations, or cooperatives to establish forest plantations in production zones of state forests.	• Rights recognized include use and management rights of timber (in production forests) and NTFPs (in production and protection forests) • Forest user groups are required to form organizations, although permits are granted at the individual (family) level. • Duration of rights is up to 35 years, after a 5-year probation period, but state retains alienation rights and rights to subsoil (minerals and oil).	**Level 2 and 3. HKM[a] and HTR[a]** allow for payments and compensation for maintaining conditions of forest landscape. **Level 4**. Customary forests, Tenure conditional of ES maintenance, reduction of land use conflict and avoided collateral damage to ES provision

Sources: DGM (2019), Monterroso et al. (2017), Siscawati et al. (2017), Monterroso and Larson (2018) and Sunderlin et al. (2018)

[a]HKM *Hutan Kemasyarakatan* community forests; HTR *Hutan Tanaman Rakyat* community forest plantations

The low-cost approach to titling promoted by Saweto DGM relies on the participation of Indigenous communities along with their regional and national federations, subnational governments, and NGOs to achieve a more efficient implementation process with greater buy-in from stakeholders. Involving stakeholders during mapping and demarcation activities can reduce conflicts and help avoid negative incentives that favor forest conversion while improving livelihoods (Blackman et al., 2017b). By 2018, DGM Saweto had reported the legal recognition of 133 new communities—a pre-condition of titling—in around 400,000 hectares (MDE Saweto Peru, 2021). More than 200 communities are expected to participate in new titling processes over the next few years with the potential to formalize up to one million hectares in key REDD+ areas. This could become a model for other countries in the Amazon Basin where legal recognition and titling of Indigenous communities has been promoted both to promote participation of those communities and to increase the effectiveness of REDD+ (Loaiza et al., 2016; Schroeder & González, 2019).

The DGM in Indonesia also aims at improving clarity and security of rights of Indigenous Peoples by supporting their recognition under Indonesia Village Law, for example, by mapping forests and village boundaries (DGM, 2022). Rights to land and forests within their ancestral territories can be recognized as "customary titled forest" (*masyarakat hukum adat*). These reforms started after the constitutional reforms in 2012 (Constitutional Court Ruling 35/PUU-X/2012). However, implementation has been slow due to lack of clear procedures and coordination of responsible government institutions (Myers et al., 2017).

The DGM in Indonesia has made more progress with social forestry schemes in state forests, which represent around 70% of Indonesia's territory. These recognize local communities' management rights and thus position them to participate in REDD+ or other PES, which could in turn both increase the value of natural resources and enhance their ability to enforce property rights (Engel & Palmer, 2008; Resosudarmo et al., 2014; Suyantoi, 2007). The DGM facilitates access to social forestry permits for both community forests (*Hutan Kemasyarakatan* HKM) and community plantations (*Hutan Tanaman Rakyat* HTR) (Krishna et al., 2017; Resosudarmo et al., 2014). HKM permits are granted to organized groups around state forests mainly for use and extraction rights, while HTR permits grant rights to state forest lands for reforestation activities. Implementation of these social forestry schemes started after the decentralization of the forest sector and reforms to the National Forest Law (Forest Law No. 41, 1999) (Siscawati et al., 2017; Banjade et al., 2016).

According to Kerr et al. (2014), these types of social forestry schemes use clarification of rights as a type of reward for environmental services. In the case of HKM, permits are granted initially for a period of 5 years, which can be extended to 25 or 35 years if communities have met their obligations. For example, organized groups of farmers may be granted tenure rights over state land in exchange for protecting forest and watershed services (Catacutan, 2011). The rights granted under HTR differ in that they allow for planting trees such as damar or rubber, thus providing an important livelihood incentive and improving local incomes. Implementation of these social forestry schemes is advancing much faster than the recognition of customary lands (Myers et al., 2017). The explicit goals addressing livelihood concerns, as well as conditional tenure rights in social forestry schemes, seem to provide clear incentives to secure key ecosystem services at least in the medium term (Suyantoi, 2007).

Thus, the DGM tenure interventions introduce conditionality both through the tenure instruments themselves, for example, restrictions on alienation rights, such as the prohibition of subdividing land or selling it for some period (cf., Bruce et al., 2010), as in the DGM Saweto in Peru, and by making the extension of social forestry permits conditional on ES provision, as in the HKM and HTR programs for communities near state forests in Indonesia. The DGM also illustrates one of the key challenges of enforcing pro-forest conditionalities: inconsistent state policies and weak monitoring (Börner et al., 2017). Kerr et al. (2014) argue that less strict conditionality is often imposed when the conditional benefits are not cash payments, such as land tenure.

Successes and Pitfalls with Formalization of Tenure Rights in the Context of Conditional Incentives

The rapid uptake of PES and then REDD+ in the Global South has provided new opportunities for securing local tenure rights (Duchelle et al., 2017; Kerr et al., 2014; Larson et al., 2013; Sunderlin et al., 2018). There is important variation in both the type of interventions and the context in which they are implemented. Readiness processes have encouraged reforms in countries to clarify, secure, and guarantee tenure rights in target areas; however, the scale of projects still seems insufficient given the long-standing and large-scale needs. In Peru, DGM implementation has been matched with similar interventions that have promoted coordination and collaboration with opportunities for scaling up interventions in the medium term. However, as pointed out by others, while titling is broadly promoted as a way to clarify tenure, it does not entirely guarantee tenure security or conservation outcomes (Engel & Palmer, 2008; Holland et al., 2014; Robinson et al., 2014). Although there are some initial assessments analyzing the impacts of titling both on forest cover and on livelihood outcomes, clearly further analysis is needed (Blackman et al., 2017b; Cruz-Burga et al., 2019). Land titling can affect conservation outcomes through multiple channels, including potentially the ability to participate in PES programs.

Indonesian social forestry schemes are an interesting example of how enhancement and recognition of tenure rights to resources can raise the value of natural resources, with benefits for both local livelihoods and forest conservation. Social forestry schemes combining different types of environmental service reward mechanisms, including the recognition of tenure rights (Resosudarmo et al., 2014; Suyantoi, 2007), demonstrate how to incorporate non-cash benefits into conditionality where there are weak tenure rights (Börner et al., 2017). Both the Peru and Indonesia cases also show the importance of enforcement of exclusion rights to ensure outcomes and meet conditionality. Having the right institutional and incentive mechanisms in place and ensuring the political will and support of tenure reforms and ability of right-holders to enforce rules is key (Naughton-Treves & Wendland, 2014; Robinson et al., 2014).

Finally, both the experience of the DGM and the scientific literature show the importance of broad participation and engagement of stakeholders including local communities as well as land and forest managers (Duchelle et al., 2018; Schroeder & González, 2019). The DGM is particularly noteworthy in that it specifically encourages participation of Indigenous Peoples and local communities in REDD+ processes. This participation has perhaps encouraged the DGM to align tenure interventions with incentive-based mechanisms and thus enhance livelihoods while ensuring provision of key ecosystem services. While the DGM provides instructive examples, we recall the lessons from the literature on PES, which clearly show that the relationship between conditional incentives and tenure security is context specific and depends on the institutional details of both the existing tenure system and the conditional incentives being introduced.

References

Agrawal, A., Wollenberg, E., & Persha, L. (2014). Governing agriculture-forest landscapes to achieve climate change mitigation. *Global Environmental Change, 29*, 270–280.

Alix-Garcia, J. M., & Wolff, H. (2014). Payment for ecosystem services from forests. *Annual Review of Resource Economics, 6*(1), 361–380.

Arriagada, R. A., Sills, E. O., Pattanayak, S. K., & Ferraro, P. J. (2009). Combining qualitative and quantitative methods to evaluate participation in Costa Rica's program of payments for environmental services. *Journal of Sustainable Forestry, 28*(3–5), 343–367.

Astuti, R., & McGregor, A. (2015). Responding to the green economy: How REDD+ and the One Map Initiative are transforming forest governance in Indonesia. *Third World Quarterly, 36*(12), 2273–2293.

Banjade, M. R., Herawati, T., Liswanti, N., & Mwangi, E. (2016). Tenure reform in Indonesia: When? What? Why? *InfoBrief No. 163*. Center for International Forestry Research, Indonesia.

Blackman, A., Corral, L., Lima, E. S., & Asner, G. P. (2017a). Building the evidence base on the forest cover effects of community titling. *Proceedings of the National Academy of Sciences of the United States of America, 114*(29), E5765.

Blackman, A., Corral, L., Lima, E. S., & Asner, G. P. (2017b). Titling indigenous communities protects forests in the Peruvian Amazon. *Proceedings of the National Academy of Sciences of the United States of America, 114*(16), 4123–4128.

Blackman, A., & Veit, P. (2018). Titled Amazon indigenous communities cut forest carbon emissions. *Ecological Economics, 153*, 56–67.

Börner, J., Baylis, K., Corbera, E., Ezzine-de-Blas, D., Honey-Rosés, J., Persson, U. M., & Wunder, S. (2017). The effectiveness of payments for environmental services. *World Development, 96*, 359–374.

Börner, J., Wunder, S., Wertz-Kanounnikoff, S., Hyman, G., & Nascimento, N. (2011). REDD sticks and carrots in the Brazilian Amazon: Assessing costs and livelihood implications. *CCAFS Working Paper No. 8. CGIAR Research Program on Climate Change, Agriculture and Food Security (CCAFS).* Copenhagen, Denmark.

Bremer, L. L., Farley, K. A., & Lopez-Carr, D. (2014). What factors influence participation in payment for ecosystem services programs? An evaluation of Ecuador's SocioPáramo program. *Land Use Policy, 36*, 122–133.

Broegaard, R. B., Vongvisouk, T., & Mertz, O. (2017). Contradictory land use plans and policies in Laos: Tenure security and the threat of exclusion. *World Development, 89*, 170–183.

Bruce, J., Wendland, K., & Naughton-Treves, L. (2010). Whom to pay? Key concepts and terms regarding tenure and property rights in payment-based forest ecosystem conservation. *Land Tenure Center Policy Brief 15*. Policy Brief, 15.

Buntaine, M. T., Hamilton, S. E., & Millones, M. (2015). Titling community land to prevent deforestation: An evaluation of a best-case program in Morona-Santiago, Ecuador. *Global Environmental Change, 33*, 32–43.

Catacutan, D. (2011). Rewards for watershed services in Sumberjaya, Indonesia. Water in the Green Economy in Practice. https://www.un.org/waterforlifedecade/green_economy_2011/pdf/biodiversity_protection_indonesia.pdf

Chomba, S., Kariuki, J., Lund, J. F., & Sinclair, F. (2016). Roots of inequity: How the implementation of REDD+ reinforces past injustices. *Land Use Policy, 50*, 202–213.

Clements, T., John, A., Nielsen, K., An, D., Tan, S., & Milner-Gulland, E. J. (2010). Payments for biodiversity conservation in the context of weak institutions: Comparison of three programs from Cambodia. *Ecological Economics, 69*(6), 1283–1291.

Cruz-Burga, Z., Monterroso, I., Larson, A. M., Valencia, F., & Saldaña, J. S. (2019). The impact of formalizing rights to land and forest: Indigenous community perspectives in Madre de Dios and Loreto. *InfoBrief No. 242*. CIFOR.

The Dedicated Grant Mechanism for Indigenous Peoples & Local Communities. (2019). *Annual Report*. World Bank.

DGM. (2022). Dedicated Grant Mechanism: Strengthening Rights and Economies of *Adat* and Local Communities. Available on-line: https://www.dgmglobal.org/indonesia, last accessed June 23, 2022.

Duchelle, A. E., de Sassi, C., Jagger, P., Cromberg, M., Larson, A. M., Sunderlin, W. D., Atmadja, N., Resosudarmo, I. A. P., & Pratama, C. D. (2017). Balancing carrots and sticks in REDD+: Implications for social safeguards. *Ecology and Society, 22*(3). https://doi.org/10.5751/ES-09334-220302

Duchelle, A. E., Simonet, G., Sunderlin, W. D., & Wunder, S. (2018). What is REDD+ achieving on the ground? *Current Opinion in Environmental Sustainability, 32*, 134–140.

Engel, S., Pagiola, S., & Wunder, S. (2008). Designing payments for environmental services in theory and practice: An overview of the issues. *Ecological Economics, 65*(4), 663–674.

Engel, S., & Palmer, C. (2008). Payments for environmental services as an alternative to logging under weak property rights: The case of Indonesia. *Ecological Economics, 65*(4), 799–809.

Evans, K., Murphy, L., & de Jong, W. (2014). Global versus local narratives of REDD: A case study from Peru's Amazon. *Environmental Science & Policy, 35*, 98–108.

Ferraro, P., & Kiss, A. (2002). Direct payments to conserve biodiversity. *Science, 298*, 1718–1719.

Figueroa, F., Caro-Borrero, Á., Revollo-Fernández, D., Merino, L., Almeida-Leñero, L., Paré, L., Espinosa, D., & Mazari-Hiriart, M. (2016). "I like to conserve the forest, but I also like the cash". Socioeconomic factors influencing the motivation to be engaged in the Mexican Payment for Environmental Services Programme. *Journal of Forest Economics, 22*, 36–51.

Galik, C. S., & Jagger, P. (2015). Bundles, duties, and rights: A revised framework for analysis of natural resource property rights regimes. *Land Economics, 91*(1), 76–90.

Hejnowicz, A. P., Raffaelli, D. G., Rudd, M. A., & White, P. C. L. (2014). Evaluating the outcomes of payments for ecosystem services programmes using a capital asset framework. *Ecosystem Services, 9*, 83–97.

Holland, M. B., de Koning, F., Morales, M., Naughton-Treves, L., Robinson, B. E., & Suárez, L. (2014). Complex tenure and deforestation: Implications for conservation incentives in the Ecuadorian Amazon. *World Development, 55*, 21–36.

Holland, M. B., Jones, K. W., Naughton-Treves, L., Freire, J.-L., Morales, M., & Suárez, L. (2017). Titling land to conserve forests: The case of Cuyabeno Reserve in Ecuador. *Global Environmental Change, 44*, 27–38.

Howell, S. (2014). No RIGHTS-No REDD: Some implications of a turn towards co-benefits. *Forum for Development Studies, 41*(2), 253–272.

Johnson, M. K., Lien, A. M., Sherman, N. R., & López-Hoffman, L. (2018). Barriers to PES programs in Indigenous communities: A lesson in land tenure insecurity from the Hopi Indian reservation. *Ecosystem Services, 32*, 62–69.

Jones, I. J., MacDonald, A. J., Hopkins, S. R., Lund, A. J., Liu, Z. Y. C., Fawzi, N. I., Blundell, A. G., et al. (2020). Improving rural health care reduces illegal logging and conserves carbon in a tropical forest. *Proceedings of the National Academy of Sciences, 117*(45), 28515–28524.

Jones, K. W., Etchart, N., Holland, M., Naughton-Treves, L., & Arriagada, R. (2020). The impact of paying for forest conservation on perceived tenure security in Ecuador. *Conservation Letters, 13*(4), e12710.

Kerr, J. M., Vardhan, M., & Jindal, R. (2014). Incentives, conditionality and collective action in payment for environmental services. *International Journal of the Commons, 8*(2), 595–616.

Krishna, V. V., Kubitza, C., Pascual, U., & Qaim, M. (2017). Land markets, property rights, and deforestation: Insights from Indonesia. *World Development, 99*, 335–349.

Larson, A. M., Brockhaus, M., Sunderlin, W. D., Duchelle, A., Babon, A., Dokken, T., Pham, T., Resosudarmo, I. A. P., Selaya, G., Awono, A., & Huynh, T.-B. (2013). Land tenure and REDD+: The good, the bad and the ugly. *Global Environmental Change, 23*(3), 678–689.

Liu, Z., Gong, Y., & Kontoleon, A. (2018). How do payments for environmental services affect land tenure? Theory and evidence from China. *Ecological Economics, 144*, 195–213.

Loaiza, T., Nehren, U., & Gerold, G. (2016). REDD+ implementation in the Ecuadorian Amazon: Why land configuration and common-pool resources management matter. *Forest Policy and Economics, 70*, 67–79.

MDE Saweto Peru. (2021). Memoria Viva. 76 pp. Available on-line: http://www.mdesawetoperu.org/sitio/wp-content/uploads/2021/08/mde_saweto_memoria_viva.pdf, last accessed June 23, 2022.

Miranda, M., Porras, I. T., & Moreno, M. L. (2003). The social impacts of payments for environmental services in Costa Rica. A quantitative field survey and analysis of the Virilla watershed. International Institute for Environment and Development, London.

Monterroso, I., Cronkleton, P., & Larson, A. M. (2019). Commons, indigenous rights, and governance. In B. Hudson, J. Rosenbloom, & D. Cole (Eds.), *Routledge handbook of the study of the commons*. Routledge.

Monterroso, I., Cronkleton, P., Pinedo, D., & Larson, A. M. (2017). Reclaiming collective rights: Land and forest tenure reforms in Peru (1960–2016). *Working Paper No. 224*. CIFOR.

Monterroso, I., & Larson, A. M. (2018). Progress in formalizing "native community" rights in the Peruvian Amazon (2014–2018). *InfoBrief No. 233*. CIFOR.

Mulyani, M., & Jepson, P. (2016). Does the 'One Map Initiative' represent a new path for forest mapping in Indonesia? Assessing the contribution of the REDD+ initiative in effecting forest governance reform. *Forests, 8*(1), 14.

Myers, R., Intarini, D., Sirait, M. T., & Maryudi, A. (2017). Claiming the forest: Inclusions and exclusions under Indonesia's 'new' forest policies on customary forests. *Land Use Policy, 66*, 205–213.

Myers, R., Larson, A. M., Ravikumar, A., Kowler, L. F., Yang, A., & Trench, T. (2018). Messiness of forest governance: How technical approaches suppress politics in REDD+ and conservation projects. *Global Environmental Change, 50*, 314–324.

Naughton-Treves, L., & Wendland, K. (2014). Land tenure and tropical forest carbon management. *World Development, 55*, 1–6.

Nelson, A., & Chomitz, K. M. (2011). Effectiveness of strict vs. multiple use protected areas in reducing tropical forest fires: A global analysis using matching methods. *PloS One, 6*(8), e22722.

Nepstad, D., Schwartzman, S., Bamberger, B., Santilli, M., Ray, D., Schlesinger, P., Rolla, A., et al. (2006). Inhibition of Amazon deforestation and fire by parks and indigenous lands. *Conservation Biology, 20*(1), 65–73.

Newton, P., Nichols, E. S., Endo, W., & Peres, C. A. (2012). Consequences of actor level livelihood heterogeneity for additionality in a tropical forest payment for environmental services programme with an undifferentiated reward structure. *Global Environmental Change, 22*(1), 127–136.

Resosudarmo, I. A. P., Atmadja, S., Ekaputri, A. D., Intarini, D. Y., Indriatmoko, Y., & Astri, P. (2014). Does tenure security lead to REDD+ project effectiveness? Reflections from five emerging sites in Indonesia. *World Development, 55*, 68–83.

Rights and Resources Initiative (RRI). (2015). *Who owns the world's land? A global baseline of formally recognized indigenous community land rights*. Rights and Resources Initiative.

Robinson, B. E., Holland, M. B., & Naughton-Treves, L. (2014). Does secure land tenure save forests? A meta-analysis of the relationship between land tenure and tropical deforestation. *Global Environmental Change, 29*, 281–293.

Robinson, B. E., Holland, M. B., & Naughton-Treves, L. (2017). Community land titles alone will not protect forests. *Proceedings of the National Academy of Sciences, 114*(29), E5764. https://doi.org/10.1073/pnas.1707787114

Robinson, B. E., Masuda, Y. J., Kelly, A., Holland, M. B., Bedford, C., Childress, Fletschner, D., Game, E., Ginsburg, C., Hilhorst, T., Lawry, S., Miteva, D., Musengezi, J., Naughton-Treves, L., Nolte, C., Sunderlin, W., & Veit, P. (2018). Incorporating land tenure security into conservation. *Conservation Letters, 11*(2). https://doi.org/10.1111/conl.12383

Rosales, R. M. P. (2003). Developing Pro-poor markets for environmental services in the Philippines (Issue 3). International Institute for Environment and Development, London.

Samii, C., Lisiecki, M., Kulkarni, P., Paler, L., & Chavis, L. (2014). Effects of payment for environmental services (PES) on deforestation and poverty in low and middle income countries: A systematic review. *Campbell Systematic Reviews, 10*(11), 1–95.

Samndong, R. A., & Vatn, A. (2018). Competing tenures: Implications for REDD+ in the Democratic Republic of Congo. *Forests, 9*(11), 662.

Schroeder, H., & González, P. N. C. (2019). Bridging knowledge divides: The case of indigenous ontologies of territoriality and REDD+. *Forest Policy and Economics, 100*, 198–206.

Simonet, G., Agrawal, A., Bénédet, F., Cromberg, M., de Perthuis, C., Haggard, D., et al. (2018). *ID-RECCO, international database on REDD+ projects and programs, linking economic, carbon and communities Data. Version 3.0.* Available online at: http://www.reddprojectsdatabase.org

Siscawati, M., Banjade, M. R., Liswanti, N., Herawati, T., Mwangi, E., Wulandari, C., Tjoa, M., & Silaya, T. (2017). *Overview of forest tenure reforms in Indonesia* (Vol. 223). CIFOR.

Smith, D. A., Holland, M. B., Michon, A., Ibáñez, A., & Herrera, F. (2017). The hidden layer of indigenous land tenure: Informal forest ownership and its implications for forest use and conservation in Panama's largest collective territory. *International Forestry Review, 19*(4), 478–494.

Sunderlin, W. D., de Sassi, C., Sills, E. O., Duchelle, A. E., Larson, A. M., Resosudarmo, I. A. P., Awono, A., Kweka, L., & Huynh, T. B. (2018). Creating an appropriate tenure foundation for REDD+: The record to date and prospects for the future. *World Development, 106*, 376–392.

Sunderlin, W. D., Ekaputri, A. D., Sills, E. O., Duchelle, A. E., Kweka, D., Diprose, R., Doggart, N., Ball, S., Lima, R., Enright, A., Hartanto, H., & Toniolo, A. (2014). *The challenge of establishing REDD+ on the ground: Insights from 23 subnational initiatives in six countries* (Vol. 104). CIFOR.

Sunderlin, W. D., Larson, A. M., Duchelle, A. E., Resosudarmo, I. A. P., Huynh, T. B., Awono, A., & Dokken, T. (2014). How are REDD+ proponents addressing tenure problems? Evidence from Brazil, Cameroon, Tanzania, Indonesia, and Vietnam. *World Development, 55*, 37–52.

Suyantoi, S. (2007). Lessons on the conditional tenure and RiverCare schemes in Sumberjaya, Indonesia: Conditionality in payment for environmental services. http://www.worldagroforestry.org/downloads/Publications/PDFS/BC07202.pdf

Swallow, B., & Meinzen-Dick, R. (2009). Payment for environmental services: Interactions with property rights and collective action. In M. Padmanabhan & V. Beckmann (Eds.), *Institutions and sustainability*. Springer.

van Noordwijk, M., & Leimona, B. (2010). Principles for fairness and efficiency in enhancing environmental services in Asia: Payments, compensation, or co-investment? *Ecology and Society, 15*(4). https://doi.org/10.5751/ES-03664-150417

Wunder, S. (2013). When payments for environmental services will work for conservation. *Conservation Letters, 6*(4), 230–237.

Wunder, S., Duchelle, A. E., Sassi Cd, S. E., Simonet, G., & Sunderlin, W. D. (2020). REDD+ in theory and practice: How lessons from local projects can inform Jurisdictional approaches. *Frontiers in Forests and Global Change, 3*(11). https://doi.org/10.3389/ffgc.2020.00011

Wunder, S., Engel, S., & Pagiola, S. (2008). Taking stock: A comparative analysis of payments for environmental services programs in developed and developing countries. *Ecological Economics, 65*(4), 834–852.

11

Strategies for Securing Tenure: The Promise and Pitfalls of Formalization

Margaret B. Holland and Moustapha Diop

What Is Land Formalization and How Does It Connect with Tenure Security?

Within the toolbox of strategies for strengthening land tenure security, the most utilized tool has been land formalization. In its simplest form, land formalization refers to a set of processes through which the state legally recognizes the rights a landholder has to property that they have held or used "without such recognition, creating new capacities and opportunities (and perhaps risks) for the right holder" (Bruce, 2012, p. 39). Formalization is often further simplified and referred to as titling or documenting property rights in a land registry or cadaster. But in practice land formalization is a more involved process through which the state formally allocates rights to landholders, whether this be recognition of previous *de facto* regimes (see Chap. 3), or through an allocation or redistribution of land to populations without title or documentation.

M. B. Holland (✉) • M. Diop
Department of Geography & Environmental Systems,
University of Maryland, Baltimore County, Baltimore, MD, USA
e-mail: mholland@umbc.edu

M. B. Holland et al. (eds.), *Land Tenure Security and Sustainable Development*,
https://doi.org/10.1007/978-3-030-81881-4_11

In this chapter, we discuss how land formalization has developed into the primary mechanism for strengthening land tenure security, initially through an exploration of post-colonial and post-independence large-scale land titling efforts. We draw on examples from Africa and Latin America to highlight some promises and pitfalls experienced in these titling campaigns. We then explore some common assumptions tied to land formalization, its relationship with tenure security, and the evolution of land formalization to its current position on global sustainability agendas. Finally, we discuss the newest generation of efforts to develop more geographically targeted approaches to land formalization, often focusing on the lands of Indigenous Peoples and local communities, to increase tenure security, improve livelihoods, and safeguard ecosystems. While the formal recognition of land rights is a necessary process tethered to promoting human rights, poverty alleviation, and sustainable development worldwide, the process of formalization is complex, political, and contested. It is in the design and implementation of land formalization that we also see a critical need for assessing existing strategies to ensure the process and outcomes are inclusive, just, and sustainable. In essence, we see formalization as remaining an important component of the land tenure security toolbox, but the way it is constructed and the techniques for using it require constant reassessment and innovation.

What Do We Mean by Land Formalization?

Property rights do not have to be formalized to exist and be legitimate (see Chap. 3). It is useful to think about property rights as fundamentally concerning "relations *between* people with regard to a thing" (Meinzen-Dick & Mwangi, 2009, p. 36). With this holistic vision of property rights, Meinzen-Dick and Mwangi (2009) go on to characterize them as overlapping bundles that operate between people in their interactions with a given resource, or between people and the land. This means that on any piece of land, there could be different rights connected to specific resources (e.g., trees, soil, wildlife, and minerals) that are held by different individuals or groups of rightsholders. Meinzen-Dick and Mwangi (2009) call this a "web of interests", which can exist and be dynamically connected to any piece of land. They refer to the process of land or tenure formalization as one of identifying interests, adjudicating them, and registering them. The institutions and governance structures that play the role of assuring and upholding rights are not necessarily always the state (Meinzen-Dick & Mwangi, 2009).

There is a tendency to further simplify and equate land formalization with land titling, when in effect land titling constitutes a typical component or step in the overall process of formalizing land tenure. Bruce (2012, p. 39) outlines land formalization as a three-phase process, carried out by the state:

1. Create a law for property rights to exist, and set a framework for the rights to be upheld;
2. Realize the rights through "titling";
3. Create an official public record of the rights through documentation/registration.

Within this framework, land titling sits squarely in the middle of a process that involves legislative, institutional, and bureaucratic actions at multiple stages. All too often, the process of land formalization stalls. This might occur at the stage of titling land where the time, money, and labor costs can be high, and conflict resolution can become more fraught than anticipated. Emerging technologies have helped streamline the process of mapping, demarcating property boundaries, and registering titles. But even these advances can be encumbered by issues related to capacity gaps, equipment maintenance, lack of reliable maps, challenges tied to land claims and disputes, or even simply access to remote rural areas.

As we will discuss in more detail, earlier large-scale land formalization efforts in the 1960s–1980s in the Global South emphasized western-based notions of individual or household titling of lands into the hands of individuals or single rightsholders, disrupting complex and overlapping layers of rights that originally existed. With the publication in 2000 of Hernando de Soto's *Mystery of Capital*, the primacy of individual freehold ownership gained further traction and a boost within the development community (Bruce, 2012). This effectively benefited some landholders, while isolating or exacerbating tenure insecurity for others.

Only relatively recently have land formalization processes focused on recognition of community-held lands and resources, or customary forms of tenure. A recent study on the process of formalization of community lands indicated that even though more than half of global land is held by communities, only 10% of that land is legally recognized by governments (RRI, 2018). Even less of that land has proceeded through all three steps to reach the point of registration and documentation in the public record (RRI, 2018; Notess et al., 2020).

What Do We Know About How Formalization of Tenure Relates to Tenure Security?

Within the context of the Sustainable Development Goals (SDGs), there has been substantial effort to recognize that formalization and tenure security are associated, and progress toward both is credited toward achieving the very first of the SDGs: "to end poverty in all its forms, everywhere" (United Nations, 2015). Land formalization is a process that involves regulatory recognition of tenure form. Tenure security is a conditional state and relates more to how and whether there is confidence rights (whether *de facto* or *de jure*) will be upheld by society (see definitions explained in Chap. 1). That said, there is a pervasive conviction and generalization that land formalization *is equal to* land titling and that a land title alone is sufficient to bring tenure security to the landholder. This is quite common among those who implement land formalization programs, who connect tenure security for landholders to the acquisition of *de jure* or statutory land title (Masuda et al., 2020). This perception stands in contrast to the community of land tenure scholars who suggest this simplified equation (formalization = land titling = tenure security) is a potentially risky and false assumption, and suggest more value be placed in assessing landholders' perceptions of tenure security (Masuda et al., 2020). As already noted, the tendency to equate land formalization with titling *only* glosses over important policy and process steps that are fundamental to the legitimacy of the land titles: the laws that define the bundle of rights tied to land, the register or cadaster that records the title and allows for updates in situations of transferal, and the capacity of the government and governance structure to be reliable and accountable to adjudicate these rights.

In many situations, land formalization has resulted in the *disruption* of tenure security for landholders. Formalization through state-led land reform, in some cases, has resulted in the erasure of communal and customary tenure in favor of private landownership (Meinzen-Dick & Mwangi, 2009; Bruce, 2012; Peluso et al., 2013). This triggers a disruption in the social fabric of community relations tied to land and shifts the context of ownership to something unfamiliar and typically unsupported in pre-existing community structures. As Bromley (2009) describes: "Titles are symbols of ownership…ownership is both a social fact and a social idea" (pp. 20–21). Some suggest the relationship between formalization and tenure security depends on who holds power and what their motivations are in driving the process of formalization. As Putzel et al. (2015) suggest, in more contemporary cases, we still see instances where the process is "top-down", driven by the state as influenced by global institutions and typically financed by multilateral or bilateral aid organizations. Often the motivation for formalization is to achieve a range of policy goals (e.g., land and resource development, poverty alleviation, growth of the tax base, and meeting the SDGs). Such cases might be less connected with local interests or realities, and therefore hold a less predictable or clear connection with increased tenure security.

In cases where land formalization has represented the overlay of new tenure rules and institutions, rather than the formal recognition of existing tenure norms and local institutions, the shift can either directly trigger conflict or result in greater tenure insecurity, most often resulting in negative impacts for the most marginalized and vulnerable populations (Platteau, 1996). Meinzen-Dick and Mwangi (2009) described this approach of simplifying land formalization and assigning the concept of ownership to a single rightsholder as slicing into the "web of overlapping interests" (p. 38). This simplification could render others without traditional rights, such as women who might traverse a piece of land to access water or harvest non-timber forest products during certain times of the year (Meinzen-Dick & Mwangi, 2009) (see additional gender-specific aspects of tenure security in Chap. 5 of this volume). Alternatively, present-day initiatives seeking formalization can originate from mobilization at a grassroots level, from communities or collective groups who "understand a need to develop or operationalize governance mechanisms in their own interests, for example, to protect a common pool or private resource or to prevent conflict" (Putzel et al., 2015, p. 455). In these situations, we might expect a clearer, more direct, and positive connection between land formalization and tenure security. And yet, even if the process is locally driven or initiated ("bottom-up"), it still requires negotiations with formal institutions and governance structures for the translation of *de facto* or customary tenure systems into that which is *de jure* and formally recognized (as explained in Chap. 3). Communities or other local landholders might already have strong tenure security prior to the formalization process. If this process is inclusive, equitable, and just, then we might expect communities to feel an increase (or no net change) in their sense of tenure security. If, on the other hand, the process fails to reflect one that is transparent and truly collaborative, even when it was driven from the bottom-up, the effect on tenure security could ultimately be to weaken it.

Historical Approaches to Formalization in Latin America

Large-scale land reform took off during the post-World War II (post-WWII) era in Latin America and in the post-independence era across much of sub-Saharan Africa.[1] Most of these operations were national-scale campaigns, often state-led (national government) and financed by multilateral development banks, labeled as agrarian or land reform, and intended to be redistributive in nature. In a critique on large-scale land reform, Bromley (2009) highlighted how these strategies were promoted

[1] We focus our review in these next sections on these two regions of the Global South, as our own research and experience extend into countries within these regions. Our review of the literature suggests that the earlier history of formalization in Asia (including South East Asia) holds different layers of complexity due to more extensive communist influence on land reform. But more recent waves of land formalization and tenure reform hold parallels with examples from sub-Saharan Africa (with the exception of China). Peluso et al. (2013) present a review of land formalization for several cases in Africa and Asia, should readers wish to extend their understanding of these comparisons.

by the Global North and imposed on the Global South under the claim that it would promote economic development and increase agricultural production. The land reform strategies set goals that mirrored the property rights regimes of global economic and colonial powers like the United States and Europe, adopting a logic that private property rights through land titling would open up a stream of economic benefits, thus securing tenure and lifting the landowner out of poverty (Bromley, 2009).

In the case of Latin America, governments began to engage in large-scale land reform starting in the early 1960s, often characterizing land reform as a way to break up large estates (*latifundia*), the remaining relics of colonial power, and get land into the hands of the landless peasantry. Often the unstated objectives were to settle regions seen as the "frontier" (e.g., the Amazon basin of South America) or stabilize regions that where international boundaries were under dispute, help to lower population and land pressures in already urbanizing or upland areas (e.g., the Andes), and to assert greater control over the poor populations (De Janvry & Sadoulet, 1989). This land reform process and promotion of frontier settlement ignored or actively erased the pre-existing territorial claims and presence of Indigenous communities in the targeted settlement regions. Reforms and land laws from this earlier generation of land formalization efforts in Latin America (1960s–1980s) tended to hold a singular focus on titling land into the hands of individual landowners, absent recognition of communal or customary forms of tenure.

State-Led Land Reform and Its Unrealized Promise in Latin America: Example of Ecuador

In 1964, Ecuador implemented a first agrarian reform with the passage of a law (updated in 1973) which labeled large regions of the country as "unsettled" (*tierras baldías*), signaling a formal erasure of the ancestral territories and active presence of multiple Indigenous groups (Bremner & Lu, 2006). The law explicitly incentivized smallholder farmers from the coastal and highland regions of the country to migrate to the Amazon and stake claim over individual plots of land. The agrarian law promised newly arriving colonists to these lands formal governmental title over a forty-hectare plot if they "improved" it (i.e., cleared the forest to cultivate), formed pre-cooperative organizations with other local farmers, and could show continued productive use of the land (Holland et al., 2014, 2017). This echoed the westward expansion and state-promoted land acquisition tied to the Homestead Act in the United States during the mid-late 1800s (De Janvry & Sadoulet, 2001).

Ultimately, title was formalized for barely one-half of those settlers who were lured by the promise of land and followed the legal guidance. For decades, thousands of families migrated there as colonists and lived there without formalized title. Indigenous Peoples across the Ecuadorian Amazon remained without any statutory recognition of their lands by the state until after the passage of the new Constitution in 2008 and Ecuador's signing of the UN Declaration on the Rights of Indigenous Peoples (UNDRIP, 2009). In fact, the new Constitution made Ecuador the first country to officially grant legal recognition and rights to nature, something we will discuss toward the end of this chapter.

Although details are different, similar types of land reform efforts were carried out across most Latin American nations during multiple decades of attempted large-scale land reform. These efforts were costly and were often deemed *inconclusive* and *incomplete* (De Janvry & Sadoulet, 2001). Land reform was inconclusive because, while it may have resulted in the awarding of statutory land title, it lacked any follow-up from the state in terms of access to services (education, health, financial) and infrastructure in often remote regions, leaving those who had migrated for land and title without any access to supports that are also critical to alleviating poverty. Land reforms were incomplete because so many failed to even formalize title for thousands of landholders, leaving them without the capital to access credit or other forms of assistance on their own (De Janvry & Sadoulet, 2001).

State-led and large-scale land reforms in Latin America were largely successful in realizing the passage of national land or agrarian reform laws—the first step in the land formalization process outlined by Bruce (2012). These laws, however, remained influenced by principles for economic development promoted by the Global North through multilateral aid organizations, where private property rights were held as highest value. With few exceptions (e.g., Mexico's land reform process from 1917 to 1992 that allowed for formalization of communal tenure as *ejidos*) (Perramond, 2008), these early-stage land laws from the 1960s to the 1980s left communal and customary forms of tenure solidly within a realm of informality, with no space for statutory recognition. In fact, a 1975 World Bank land reform policy brief firmly recommended that these tenure systems be abandoned by governments in favor of individual "freehold" title and privatization of common-pool resources (Deininger & Binswanger, 1999). Later decades would see this perspective solidly refuted through the dedicated work of scholars like Dr. Elinor Ostrom (e.g., Schlager & Ostrom, 1992), indicating not only the disruptions to tenure security that can occur when land privatization cuts through complex tenure systems or a "web of interests" (Meinzen-Dick & Mwangi, 2009), but also the disruption and fragmentation this can trigger for ecosystems (Boyd et al., 2018).

Evolving Generations of Land Reform and Formalization in Africa

In a similar way to Latin America, the nations of sub-Saharan Africa embarked on a wave of large-scale, state-led land reforms starting in 1960, as each nation was achieving independence from colonial powers. Since the beginning of these land reform processes, governments seemed to hold two conflicting goals: (1) to formally recognize the rights of traditional landholders across rural regions to strengthen tenure security, and (2) to formalize their tenure systems to attract foreign investment and boost agricultural production (Chimhowu, 2019; Diop, 2020). Even in a state of newly formed independence, many young African nations, especially those emerging from former French or British rule, tended to follow an approach to land reform, formalization, and administration that aligned with the example of their colonizers, with top-down strategies for implementation of large-scale land reform, and often classifying large swaths of the rural landscape as "open land". This first generation of large-scale land reform in Africa experienced similar setbacks and pitfalls to those in Latin America: expensive programs that were bureaucratically burdensome, centralized, and top-down, and ultimately reforms that were left incomplete. The approach by the newly formed governments was often to ignore customary tenure systems or actively make them illegitimate according to state law. The result of these early formalization programs was that they typically exacerbated tenure insecurity, conflict, and inequality among local communities (Diop, 2020).

Land Reform and Formalization in the Era of Decentralization

The second wave of land reform began during the 1990s, with a goal of bringing adjudication powers closer to local people by delegating power from the central governments to local government authorities and elected officials. Here, the assumption was that decentralized agents could more efficiently and accurately meet the needs of local communities (Diop, 2020). This drive for decentralization of land formalization was both in response to the failures of the first generations of large-scale state-led land reforms, but also aligned with parallel efforts to decentralize natural resource and forest management in regions across the Global South, initially led in different parts of South and Southeast Asia (Agrawal & Ostrom, 2001; Balooni & Inoue, 2007).

Perhaps the biggest achievements from these second generation of land reform and formalization programs were that new land laws passed during the 1990s and early 2000s (e.g., South Africa, Uganda, Kenya, and Mozambique) recognized customary tenure rights and allowed for the issuance of customary title (Wily, 2018). This wave of legislative reform and recognition of customary rights swept through 39 out of 54 countries in sub-Saharan Africa, as documented for 1990–2017 (Chimhowu, 2019). During this same time period, multilateral and bilateral aid organizations began to promote both decentralization and the formal recognition of community lands. Even the World Bank, one of the most prominent champions of large-scale land reforms and private land formalization, supported prioritizing communal tenure systems given evidence from early studies on tenure security that communal tenure can represent increased tenure security and lower transaction costs than freehold or individual title (Deininger & Binswanger, 1999). Lawry et al.

(2017) would later go on to suggest that formalizing customary tenure may not increase tenure security in the region. He and other researchers term this the "Africa effect", where local landholders already perceived their security of tenure in customary systems to be high prior to formalization, resulting in little-to-no measurable effect of tenure formalization on tenure security.

Despite the passage of progressive land legislation across multiple countries, one principal concern of the decentralized and "bottom-up" approach to formalization was that the formal recognition of existing customary structures would result in exacerbating intra-community challenges and inequitable power structures (e.g., further advantaging elite members of the community), effectively locking in these disparities, thus risking further marginalization of specific population groups (e.g., women, religious minorities, and recent migrants). In this way, the shift to more locally driven formalization often solidified and entrenched such power disparities. Impact evaluation studies confirmed this across multiple contexts (see, e.g., Ribot, 2009; Ghebru & Lambrecht, 2017), revealing that these programs benefited local educated and elite populations and widened the gender equity gap (Diop, 2020).

Ultimately, the second generation of land reform still suffered setbacks, although there are bright spots (e.g., Chap. 12). Also, despite the progressive legislation and the decision of many governments to recognize (formalize) customary tenure, very few made any steps to implement these legal frameworks, and across the continent of Africa there remains a large area (an estimated 78% of arable land) that is without formal title and still primarily under customary tenure systems (Alden Wily, 2018).

A New Wave of Land Formalization in Sub-Saharan Africa and Latin America

Over the past fifteen years, we have seen new variations of land formalization efforts take hold in various countries across Latin America and sub-Saharan Africa. Nations passed new legislation formally recognizing the rights of those holding customary tenure, as already noted for the majority of nations in the African continent, and for Indigenous Peoples and

Afro-descendant communities in Latin America (see Chap. 4), marking a significant increase in the amount of land under customary and communal tenure gaining statutory recognition (RRI, 2018; Chimhowu, 2019).

There is still top-down and bottom-up initiation of land formalization, and the current era signals a mixing of the two. Several countries in sub-Saharan Africa implemented large-scale land regularization programs (e.g., Rwanda and Ethiopia), with the objectives of improving land registration, increasing agricultural production, and ultimately strengthening tenure security (Bizoza & Opio-Omoding, 2021). Titling programs in this recent wave of land formalization have been more targeted geographically, often involving pilot project phases for ongoing impact evaluation, assessment, and adaptation.

Land titling, mapping, and registration efforts have surged once again, after a relative decline in investment during the 1990s and early 2000s. In the post-2000 era, new international funds and institutions have formed to help finance land formalization and strengthen land tenure security. In addition to the original set of organizations financing and promoting land formalization (e.g., World Bank and regional development banks, the European Union, and the Food & Agricultural Organization [FAO] of the United Nations), we have seen a diversification in the donor landscape with the creation of dedicated funds from bilateral donors like Germany (GIZ), France (AfD), and Sweden (SIDA), as well as the Millennium Challenge Corporation (MCC) in the United States. Since 2005 alone, there have been 98 projects financed by these entities, specifically involving land titling and registration across 29 countries, and totaling around 770 million USD (D-Portal, 2021).

Technological innovation and advancement have enabled the creation of sophisticated digital platforms for land delineation and registration, which have promised to speed up the process, increase transparency and integrity of these systems, allowing for updates of land records. The newer generations of land titling and registration, now more contemporarily referred to as land regularization, have also involved emerging technologies—though these have faced challenges in terms of technology access, expense, and upkeep of registration systems. At times, these precision technologies, particularly in mapping and delineating property boundaries, have brought with them new instances of boundary disputes and conflict. They have also on occasion caused some hesitation and resistance from communities who worried about having their precise land boundaries suddenly "visible" to the state. In the case of mapping and registering of Indigenous territories in Latin America, for example, the map typically only reflects the territorial boundary, or "tenurial shell" (Barry & Meinzen-Dick, 2008; Smith et al., 2017), and does not include more layered ways in which communities might hold tenure rights over other resources, allow for individual rights within communal territories, or access rights to other lands. At the same time, the strategy of formalizing recognition of the overall territory allows for a more efficient process and engagement between the community and the state, and leaves flexibility for the layers of tenure within the community boundary to adjust and update over time. Importantly, despite these technological advances, such tools cannot substitute the deliberate and important work of inclusive and collaborative titling and mapping with individuals and communities, which ends up taking more time.

Converging Drivers of Formalization and New Approaches

The trends during the first two decades of the millennium indicate that land formalization has maintained a central role in the pursuit of global goals tied to poverty alleviation and sustainable development. While a strong emphasis on formal recognition of full property rights for individuals persists, the past decade or more has additionally focused on formal recognition of lands held by customary and communal tenure systems. When examining the motivations and drivers behind this uptick in formalization for communities, we see initiatives that originate from both top-down and bottom-up mobilizations to formalize. These reflect social and environmental goals tied to sustainable development objectives as well as social, environmental, and climate justice concerns.

Formalization for Community Empowerment or Dispossession?

The newer surge in formalization in many cases reflects newer and increasing pressures on land and resources for productive or extractive uses. As reviewed in the chapter on land grabbing and large-scale land investments (Chap. 7), we have seen examples of what Maganga et al. (2016) refer to as "dispossession through formalization" in countries like Tanzania and Mozambique. In such cases, progressive land laws have set the stage for formal recognition of customary tenure, and outside investor interest has influenced the state in prioritizing certain lands for formalization. This has set the stage for investors to negotiate lease access directly and immediately with communities upon reaching formal title and registration status. The weakness of the state in adjudicating tenure, combined with its duality of interest in developing its economy, results in the government playing a minor role, ensuring the bare minimum of assistance and support to communities in the process. Communities find themselves in a position where they are learning about the context of their rights within the statutory framework of their newly formalized title, yet doing so from a position of less familiarity with legal systems, lower levels of education, and limited capacity to negotiate effectively. The result is that communities are often left disadvantaged and less secure, even under the veneer of above-the-board legality (Nhantumbo & Salomão, 2010).

The examples emerging from Tanzania and Mozambique point to the increased understanding that the *timing* of formalization can be as important as the *direction* from which the process is initiated (top-down vs. bottom-up) (Putzel et al., 2015). In these examples, the timing of the decision to engage in a formalization process is due to pressure and interests from outside entities, thereby setting the stage for communities to be unequal partners in the formalization process from the outset, risking increased tenure *in*security. Even when the promise of outside investment is tied to promised benefits of poverty reduction (e.g., through generating jobs and some level of benefit-sharing from the resource production or extraction), the implementation of the formalization process can often leave communities in a weakened position to negotiate these benefits, or to have confidence that negotiated rights of access, withdrawal, and compensation will be upheld (Nhantumbo & Salomão, 2010).

Formalization for Conservation and Climate Mitigation

On the other side, state interests in prioritizing formalization for development and extractive interests is a top-down approach to formalization that can seek to increase protection of ecosystems, ecological restoration, and climate change mitigation through nature-based approaches to carbon storage and sequestration. Most of these formalization efforts focus on tropical forests and forest tenure. Certain state entities (e.g., the Ministry of Environment) can join with international conservation organizations to promote targeted formalization and recognition of rights for the implementation of programs. For example, payment for ecosystem services (PES), such as Reducing Emissions from Deforestation and forest Degradation plus (REDD+), are generally predicated on clear and uncontested title before participation (Holland et al., 2017; Robinson et al., 2018). Chapter 10 explored in depth the relationship between PES, REDD+, and tenure security, as well as the critiques of these top-down approaches to formalization so that landholders may participate in such conservation incentive and benefit programs.

Regardless, in addition to formalizing land for eligibility to participate in such programs we also see an emphasis on prioritizing titling programs to reduce deforestation, even in the absence of any added conservation incentive or policy. The research evidence on whether titling alone leads to slowing forest loss is mixed and still building, with limited empirical evidence yet on long-term, sustained impacts (Holland et al., 2017; Robinson et al., 2018; Tseng et al., 2020). Two recent studies are among those that point to positive outcomes for forests when formalizing community lands in the Brazilian Amazon (Baragwanath & Bayi, 2020), and the West African nation of Benin (Wren-Lewis et al., 2020) (although BenYishay et al. (2017) find contrasting evidence). These types of rigorous evaluations on the impact of titling are lending further credence to aligning the interests of the state in land formalization that can further national commitments to achieving global goals for biodiversity conservation and climate mitigation.

Seeking Formal Recognition as a Form of Resistance to Land and Resource Development

Recent trends in formalization have also reflected processes that are initiated from the bottom-up, often reflecting landholders' collective concerns about impending pressures for access or expropriation of their lands. In late 2020, the Naso Tjër Di people of northern Panama succeeded in having a portion of their ancestral territory legally recognized by the government as a *comarca*, a form of semi-autonomous region, which the Panamanian government first began recognizing in 1938 with the recognition of the Guna Yala comarca. The Naso comarca represents only the sixth comarca, with approximately 30 Indigenous communities still seeking this level of recognition for their traditional territories (IWGIA, 2021). This recent declaration represented the culmination of a more than twenty-year struggle for formal recognition of Naso territorial rights. For years, the Naso have actively protested hydroelectric and other resource development on their lands, surrounding the Teribe River (with hydroelectric dam finalized in 2014). At the same time, their quest for recognition as a comarca was often sidelined because of the overlap between their lands and two protected areas (La Amistad Biosphere Reserve and Palo Seco Protected Forest, both UNESCO World Heritage sites) that were formally recognized in the mid-2000s, committing the state to uphold strict conservation objectives on those lands. However, evolving perspectives and appreciation for the alignment of Indigenous land management and conservation goals have resulted in the Ministry of Environment issuing a statement endorsing the formal recognition of the comarca, seeing it as complementary to the conservation goals tied to the two protected areas. This was echoed in the President of Panama's recent visit to the comarca, noting that the forests on those lands would now be "doubly-protected" (Cannon, 2021).

Even for Indigenous communities that might have achieved legal recognition automatically through legislation (eliminating the need to seek a full process of formalization), many communities are pressing for full land registration so that they can "double-lock" their rights (Alden Wily, 2017; Notess et al., 2020). These processes cost more in terms of time and investment for communities, but communities often perceive an added value in being able to delineate boundaries and make their lands clearly legible to the state and outside interests in a formal land registry. The directionality of these efforts (bottom-up) potentially points to a higher likelihood of formalization representing increased tenure security for these communities. But much of the effect on tenure security will still depend on the timing of the formalization process (including how long it takes overall), and the response by the state in upholding those rights.

Formalization as Legal Recognition of the Rights of Nature

Earlier in this chapter, we noted that Ecuador became the first country to formally recognize the Rights of Nature (*Pachamama*) in its 2008 Constitution. Since that point, Indigenous communities, municipalities, and townships across eight countries have taken steps to gain legal recognition of the Rights of Nature, either tied to "all Nature" (as in the case of Ecuador), unique ecosystems (municipalities in the United States), or specific landscape features and rivers (New Zealand) (Kauffman & Martin, 2018). While these represent efforts to achieve statutory recognition of the Rights of Nature, the attribution of rights to Nature has long been a component of customary law and tradition for Indigenous communities (Cano Pecharroman, 2018). In the case of Ecuador, Nature is recognized within the Constitution as having the rights to exist, sustain ecosystem integrity, and be restored when degraded (Kauffman & Martin, 2018). Importantly, the constitutional recognition in Ecuador grants any person the right to demand that the state uphold the rights attributed to Nature (Cano Pecharroman, 2018). We are still in the initial phase of understanding how efforts to formally recognize the Rights of Nature will interact with community-driven and bottom-up initiatives to formalize rights to land and resources, but early cases point to instances where communities have been able to assume the role of Nature's "custodian" in a legal sense, advocate for the protection of rivers running through their lands, and leverage alliances with environmental organizations for increased protection of ecosystems important to them (Cano Pecharroman, 2018).

Aligning Top-Down and Bottom-Up for Strengthening Tenure Security

The drive for land and resource formalization has gained significant ground in these recent years, responding to different pressures on landholders and to policy goals and mandates for states. In this chapter, we have reviewed the history (post-WWII) of land formalization in Latin America and sub-Saharan Africa, and how the processes of land formalization have diversified and evolved over time. Importantly, we have examined the "sticky", but ultimately misinformed, messaging practitioners and policymakers tend to adopt that equate land formalization with titling only, and with improved tenure security. Finally, we have reviewed recent trends in formalization efforts applying a dual lens of *directionality* and *timing* to shed light on how these might ultimately align to help strengthen tenure security for landholders in the Global South. With directionality, joining together top-down and bottom-up approaches will add even greater complexity to processes that can already be slow, intensive, and expensive. But this is a necessary step, and the timing of it matters for respecting the full rights of communities. Ultimately, we see success in achieving the Sustainable Development Goals as depending on this integration and alignment of formalization processes, as it requires a process of negotiated understanding between communities of people and their system of formal governance, which will only reach sustainable outcomes if that process is equitable and just.

References

Agrawal, A., & Ostrom, E. (2001). Collective action, property rights, and decentralization in resource use in India and Nepal. *Politics & Society, 29*(4), 485–514.

Alden Wily, L. (2017). Customary tenure: Remaking property for the 21st century. In *Comparative property law*. Edward Elgar Publishing.

Alden Wily, L. (2018). Collective land ownership in the 21st century: Overview of global trends. *Land, 7*(2), 68.

Balooni, K., & Inoue, M. (2007). Decentralized forest management in South and Southeast Asia. *Journal of Forestry, 105*(8), 414–420.

Baragwanath, K., & Bayi, E. (2020). Collective property rights reduce deforestation in the Brazilian Amazon. *Proceedings of the National Academy of Sciences, 117*(34), 20495–20502.

Barry, D., & Meinzen-Dick, R. (2008, July). The invisible map: Community tenure rights. In *12th Conference of the International Association for the Study of the Commons.*

BenYishay, A., Heuser, S., Runfola, D., & Trichler, R. (2017). Indigenous land rights and deforestation: Evidence from the Brazilian Amazon. *Journal of Environmental Economics and Management, 86*, 29–47.

Bizoza, A. R., & Opio-Omoding, J. (2021). Assessing the impacts of land tenure regularization: Evidence from Rwanda and Ethiopia. *Land Use Policy, 100*, 104904.

Boyd, R., Richerson, P. J., Meinzen-Dick, R., De Moor, T., Jackson, M. O., Gjerde, K. M., … Dye, C. (2018). Tragedy revisited. *Science, 362*(6420), 1236–1241.

Bremner, J., & Lu, F. (2006). Common property among indigenous peoples of the Ecuadorian Amazon. *Conservation and Society, 4*(4), 499–521.

Bromley, D. W. (2009). Formalising property relations in the developing world: The wrong prescription for the wrong malady. *Land Use Policy, 26*(1), 20–27.

Bruce, J. W. (2012). Simple solutions to complex problems: Land formalization as a 'Silver Bullet'. Chap. 2 in J. M. Otto & A. Hoekema (Eds.), *Fair land governance how to legalize land rights for rural development* (pp. 31–55). Leiden University Press.

Cannon, J. C. (2021). Indigenous community wins recognition of its land rights in Panama. Mongabay Series: Land Rights & Extractives. Retrieved May 29, 2021, from https://news.mongabay.com/2021/02/indigenous-community-wins-recognition-of-its-land-rights-in-panama/#:~:text=A%20ruling%20by%20Panama's%20Supreme,secure%20the%20Naso's%20land%20rights

Cano Pecharroman, L. (2018). Rights of nature: Rivers that can stand in court. *Resources, 7*(1), 13.

Chimhowu, A. (2019). The 'new' African customary land tenure. Characteristic, features and policy implications of a new paradigm. *Land Use Policy, 81*, 897–903.

D-Portal. (2021). Development Portal online database for IATA Registry. Retrieved April 25, 2021, from http://d-portal.org/ctrack.html#view=search

De Janvry, A., & Sadoulet, E. (1989). A study in resistance to institutional change: The lost game of Latin American land reform. *World Development, 17*(9), 1397–1407.

De Janvry, A., & Sadoulet, E. (2001, March). Access to land and land policy reforms. In A. de Janvry, G. Gordillo, J-P. Platteau, & E. Sadoulet (Eds.), *Access to land, rural poverty and public action*. UNU-WIDER, Oxford University Press.

Deininger, K., & Binswanger, H. (1999). The evolution of the World Bank's land policy: Principles, experience, and future challenges. *The World Bank Research Observer, 14*(2), 247–276.

Diop, M. (2020). To formalize or not: Investigating drivers of landholders' decisions for land formalization in rural sub-Saharan Africa, the case of Senegal. University of Maryland, Baltimore County. 206 pp.

Ghebru, H., & Lambrecht, I. (2017). Drivers of perceived land tenure (in) security: Empirical evidence from Ghana. *Land Use Policy, 66*, 293–303.

Holland, M. B., De Koning, F., Morales, M., Naughton-Treves, L., Robinson, B. E., & Suárez, L. (2014). Complex tenure and deforestation: Implications for conservation incentives in the Ecuadorian Amazon. *World Development, 55*, 21–36.

Holland, M. B., Jones, K. W., Naughton-Treves, L., Freire, J. L., Morales, M., & Suárez, L. (2017). Titling land to conserve forests: The case of Cuyabeno Reserve in Ecuador. *Global Environmental Change, 44*, 27–38.

IWGIA. (2021). The Indigenous World. International Work Group for Indigenous Affairs. 35th edition, 824 pp. www.iwgia.org

Kauffman, C. M., & Martin, P. L. (2018). Constructing rights of nature norms in the US, Ecuador, and New Zealand. *Global Environmental Politics, 18*(4), 43–62.

Maganga, F., Askew, K., Odgaard, R., & Stein, H. (2016). Dispossession through formalization: Tanzania and the G8 land agenda in Africa. *Asian Journal of African Studies, 40*, 3–49.

Masuda, Y. J., Kelly, A. C., Robinson, B. E., Holland, M. B., Bedford, C., Childress, M., Game, E. T., Ginsburg, C., Hilhorst, T., Lawry, S. W., Miteve, D. A., Musengezi, J., Naughton-Treves, L., Sunderlin, W. D., & Veit, P. (2020). How do practitioners characterize land tenure security? *Conservation Science and Practice, 2*(5), e186.

Meinzen-Dick, R., & Mwangi, E. (2009). Cutting the web of interests: Pitfalls of formalizing property rights. *Land Use Policy, 26*(1), 36–43.

Nhantumbo, I., & Salomão, A. (2010). *Biofuels, land access and rural livelihoods in Mozambique*. International Institute for Environment and Development (IIED).

Notess, L., Veit, P., Monterroso, I., Sulle, E., Larson, A. M., Gindroz, A. S., Quaedvlieg, J., & Williams, A. (2020). Community land formalization and company land acquisition procedures: A review of 33 procedures in 15 countries. *Land Use Policy*, 110, p.104461.

Peluso, N. L., Kelly, A. B., & Woods, K. (2013). Context in land matters: Access effects and history in land formalization. In *Commoners and the Changing Commons: Livelihoods, Environmental Security, and Shared Knowledge*, the Fourteenth Biennial Conference of the International Association for the Study of the Commons, Mt Fuji, Japan, 48pp.

Perramond, E. P. (2008). The rise, fall, and reconfiguration of the Mexican ejido. *Geographical Review, 98*(3), 356–371.

Platteau, J. P. (1996). The evolutionary theory of land rights as applied to sub-Saharan Africa: A critical assessment. *Development and Change, 27*(1), 29–86.

Putzel, L., Kelly, A. B., Cerutti, P. O., & Artati, Y. (2015). Formalization as development in land and natural resource policy. *Society & Natural Resources, 28*(5), 453–472.

Ribot, J. C. (2009). Authority over forests: Empowerment and subordination in Senegal's democratic decentralization. *Development and Change, 40*(1), 105–129.

Robinson, B. E., Masuda, Y. J., Kelly, A., Holland, M. B., Bedford, C., Childress, M., Fletschner, D., Game, E. T., Ginsburg, C., Hilhorst, T., Lawry, S., Miteva, D. A., Musengezi, J., Naughton-Treves, L., Nolte, C., Sunderlin, W. D., & Veit, P. (2018). Incorporating land tenure security into conservation. *Conservation Letters, 11*(2), e12383.

RRI. (2018, November). At a crossroads: Consequential trends in recognition of community-based forest tenure from 2002–2017. *Rights & Resources Initiative Report*, 60pp.

Schlager, E., & Ostrom, E. (1992). Property-rights regimes and natural resources: A conceptual analysis. *Land Economics, 68*, 249–262.

Smith, D. A., Holland, M. B., Michon, A., Ibáñez, A., & Herrera, F. (2017). The hidden layer of indigenous land tenure: Informal forest ownership and its implications for forest use and conservation in Panama's largest collective territory. *International Forestry Review, 19*(4), 478–494.

Wren-Lewis, L., Becerra-Valbuena, L., & Houngbedji, K. (2020). Formalizing land rights can reduce forest loss: Experimental evidence from Benin. *Science Advances, 6*(26), eabb6914.

United Nations. (2015). Transforming our World: The 2040 Agenda for Sustainable Development. A/RES/70/1. New York, NY. 41 pp.

Lawry, S., Samii, C., Hall, R., Leopold, A., Hornby, D., & Mtero, F. (2017). The impact of land property rights interventions on investment and agricultural productivity in developing countries: a systematic review. Journal of Development Effectiveness, 9(1), 61-81.

Tseng, T.W.J., Robinson, B.E., Bellemare, M.F., BenYishay, A., Blackman, A., Boucher, T., Childress, M., Holland, M.B., Kroeger, T., Linkow, B. and Diop, M., 2021. Influence of land tenure interventions on human well-being and environmental outcomes. Nature Sustainability, 4(3), pp.242-251.

United Nations. (2007). United Nations Declaration on the Rights of Indigenous Peoples : resolution / adopted by the General Assembly, 2 October 2007, A/RES/61/295.

12

Securing Communal Tenure Complemented by Collaborative Platforms for Improved Participatory Landscape Management and Sustainable Development: Lessons from Northern Tanzania and the Maasai Mara in Kenya

Hannah L. Timmins, Patricia Mupeta-Muyamwa, Jackson Marubu, Chira Schouten, Edward Lekaita, and Daudi Peterson

The Relationships between Tenure, Sustainable Development, and Conservation

Community-based conservation (CBC) is a concept that has been evolving since the late 1980s (Barrow & Murphree, 2001) and encompasses a wide range of approaches. CBC essentially maintains that conservation

H. L. Timmins (✉)
Independent Consultant, Nairobi, Kenya

P. Mupeta-Muyamwa • J. Marubu • C. Schouten
The Nature Conservancy, Arusha, Tanzania

M. B. Holland et al. (eds.), *Land Tenure Security and Sustainable Development*,
https://doi.org/10.1007/978-3-030-81881-4_12

should be participatory, treating local stakeholders as project partners through collective action, and ideally yield some benefits for local communities (Adams & Hulme, 2001) in a trade-off mechanism to recognize the costs of conservation.

Securing tenure for local communities is one much-utilized approach. The motivation is that securing tenure ensures access to future resources and thus invokes a sense of responsibility for managing resources sustainably, and also acts as a benefit itself and increases buy-in to facilitate greater cooperation on project rules and interventions (Brooks et al., 2013).

A systematic analysis of 136 CBC projects by Brooks et al. (2013) assessed multiple community characteristics against project goals and found secure tenure to be the characteristic most strongly associated with achieving economic goals for communities. However, much is still unknown about how secure tenure for communities contributes toward social and ecological project goals, or under what conditions this approach works best. With the rest of the world increasingly devolving land to communities in an attempt to secure conservation goals (Robinson et al., 2017), it is critical to take a closer look at the mechanisms by which this might happen and the elements that make projects successful. We present two case studies that illustrate the complexities of how CBC programs intersect with tenure security and other factors to attain project-level human well-being and conservation goals.

The Mara-Serengeti-Tarangire Ecosystem Chain: Two Case Studies across One Border

Along the Southern Kenya–Northern Tanzania border lies a chain of interconnected ecosystems: the Maasai Mara in Kenya joined to Tanzania's Serengeti ecosystem, and finally the Tarangire ecosystem to the East.

E. Lekaita
Ujamaa Community Resource Team, Arusha, Tanzania

D. Peterson
Dorobo Fund, UCRT, Dorobo Safaris, Arusha, Tanzania

Within this ecosystem chain sits the Maasai Mara National Reserve, the Serengeti National Park, the Ngorongoro National Park, the Tarangire National Park, the Loliondo Game Controlled Area, the Maswa Game Reserve, and the Lake Natron Basin Ramsar Site—creating a patchwork of protected areas. Bookmarking these protected areas are our two case studies: the Maasai Mara Wildlife Conservancies Association (MMWCA), and 230 km to the south-east, a collection of communal Certificates of Customary Rights of Occupancy (CCROs) held by the Hadzabe Indigenous Peoples.

Historically, communities in this region have endured regimes where they were disempowered from land management decisions through land grabs from colonial governments, and, following independence, top-down resource management plans from modern governments. At the community level, legislative frameworks enacted by modern government regimes have fueled further land grabs, conflicts, land conversion, fence building, and have resulted in biodiversity loss and adverse impacts on human well-being. More recently, there has been a movement toward grassroots approaches to resource management and conservation through projects underpinned by securing tenurial titles, such as Kenyan Community Conservancies[1] (Kenya's Wildlife Conservation and Management Act 2013) and Tanzanian Certificate of Customary Right of Occupancy (CCROs—Tanzania's Village Lands Act No. 51999).

These communities live within a vibrant wildlife corridor, and communities that are the target of the case studies have a history of sustainably living with wildlife. Each year, 1.3 million wildebeest, 200,000 zebra, and hundreds of thousands of gazelle travel 3000 km across this landscape in the largest and most diverse mammal migration in the world. The area provides critical habitat for lions, elephants, cheetahs, black rhinoceros, African wild dogs, and pangolins. The Maasai Mara portion of the ecosystem accounts for 25% of Kenya's wildlife population

[1] Community conservancies are established by a community on community land. Communities are the dominant decision makers and enforcers; they democratically elect a representative board from the community. Ex-officio board members from Kenya Wildlife Service, conservation and tourism partners also hold seats on the board. Sub-committees on finance, grazing, and tourism may be established to drive strategic plans and provide oversight. The Board determines benefit-sharing mechanisms, drives strategic development of the conservancy, and oversees operational management (KWCA, 2020).

(Western et al., 2009), is home to one of Kenya's last remaining viable lion populations (Schuette et al., 2013), and is designated an Important Bird Area (IBA) by BirdLife International, as it supports over 550 species of birds (BirdLife International, 2019). Meanwhile, Northern Tanzania's savanna rangelands, where the communal CCROs are situated, support approximately 4200 elephants and 35,000 migratory zebra and wildebeest (Northern Tanzania Rangelands Initiative, 2019). In this landscape, wide-ranging species require a patchwork of protected areas, corridors, stepping stones, and spill-over reserves in order to continue annual migrations and maintain genetic population health through dispersal (Breckheimer et al., 2014). The conservancies of the Mara and the communal CCROs of the Northern Tanzania rangelands aim to supplement these areas and increase landscape permeability, while also advancing human well-being goals of local communities.

The Last Vestige of the Hadzabe: Tanzania's Last Hunter-Gatherers

For around 80,000 years, the Hadzabe, Tanzania's last hunter-gatherers, have roamed the Northern Tanzania rangelands between their wet and dry season woodland refuges. Traditionally living in camps of between 10 and 50 people, the Hadzabe splinter into smaller groups to gather berries, tubers and greens, hunt game with poisoned and un-poisoned arrows, and whistle for honeyguide birds to lead them to bee nests (Redfern, 2018).

"Unlike their landscape neighbors the Maasai pastoralists, the Hadzabe live in woodlands, not on the plains," says Chira Schouten, Project Lead for the Northern Tanzania Rangelands Initiative (NTRI). "Pastoralists on the plains and plateaus will search for water and grass during the dry season, whereas the Hadzabe will use baobab trees and small springs as water sources. Occupying these different niches allowed pastoralists and hunter-gatherers to utilize different areas within the landscape in rotation, reducing competition."

Today, the last 1300 Hadzabe live in the Yaeda-Mangola landscape, occupying just 10% of their ancestral lands and only a few hundred continue their traditional hunter-gatherer lifestyle. Hadzabe and the wildlife they rely on have been squeezed onto smaller and smaller areas by conversion of lands to agriculture and increasing pressure from livestock grazing by non-Hadzabe. Wildlife has found some protection in the surrounding protected areas outside of Hadzabe lands, but like many traditional cultures, some Hadzabe have suffered a history of displacement for protected area creation and consequently view modern conservation as synonymous with land loss. As a communal culture with a light ecological footprint and a history of living outside of Tanzania's economic and political system, demarcating and securing Hadzabe homelands have been a struggle.

Conversion to Agriculture and Land Grabs: The Loss of Hadzabe Lands and Resources

During the 1990s, Tanzania's land reform laws attempted to recognize and devolve more land to communities through customary rights to use and manage lands (Ujamaa Community Resource Team, 2014). The Village Land Act No. 5 (1999), in particular, designated Village Councils (i.e., elected village leaders) as managers of Village Lands with legally registered boundaries. In turn, the Village Councils were accountable to Village Assemblies (i.e., all the adult members of a village (Williams, 2017)) for all land-use and allocation decisions.

These reforms provided a framework for establishing multiple Hadzabe villages in the Yaeda-Mangola landscape, aiming to support the preservation of their traditional lifestyles. However, in practicing traditional lifestyles, the Hadzabe leave their camps for periods after harvesting the berries, roots, and honey from the area (Northern Tanzania Rangelands Initiative, n.d.). "Meanwhile, Tanzania's rapidly increasing population and demand for land drove farmers and pastoralists into what was historically Hadzabe land," says Chira. "Rather than confront incoming farmers or pastoralists the Hadzabe simply moved away as theirs is a culture of consensus."

While the land reform laws allowed the Hadzabe to register settlements as administratively recognized villages, in practice the apolitical Hadzabe were unable to control immigration by neighboring agriculturalists and pastoralists. As a result, the landscape was soon expropriated from predominantly traditional and wild species' uses, to divided monocultures of low-yield beans, onions, maize, and grasslands by non-Hadzabe agriculturalists and pastoralists. As more non-Hadzabe moved in, this land-use change was accompanied by a shift in governance, and by 2009 the majority of Village Council representatives were non-Hadzabe, which led to further subdivision of village tenure and allocation of lands to non-Hadzabe Tanzanian nationals.

As a minority with little influence over village decisions, the Hadzabe retreated into smaller areas of land and natural habitat and many Hadzabe became dependent on supplemental maize. "Hadzabe oral history has no record of famine, which they attribute to reliance on a large diversity of adapted plants and animals rather than a few domesticated crops and animals," according to Daudi Peterson from Dorobo Tours.

Certificate of Customary Right of Occupancy: Retrofitting Communality to the Village Acts

In the 1980s, Dorobo Tours, a tourism business operating in the Yaeda-Mangola landscape, was becoming increasingly concerned with habitat fragmentation and charcoal burning in the area. They met this challenge by striking agreements with communities to provide income from tourism in exchange for setting aside areas of village land for wildlife. Eventually Dorobo Tours partnered with a few local activists to form Ujamaa Community Resource Team (UCRT) to better support communities in land-use planning and building governance capacity.

In 2005, UCRT began working with the Hadzabe and the Ministry of Lands to:

(i) develop participatory land-use plans to zone areas for wildlife, livestock, and agriculture, and reduce further settlement and subdivision;

(ii) provide legal tenure security through communal Certificate of Customary Right of Occupancy; and
(iii) develop village bylaws whereby the local government could enforce land-use plans.

The CCRO title was originally developed under Tanzania's Village Lands Act No. 5 in 1999 as the legal framework utilized by individuals living in a village to document and formalize private land. Up until this time, CCROs had only been used as an instrument to secure individual tenurial titles and had not been applied for *communal* titles to recognize customary lands. Customary lands are defined as village lands under Tanzania's Lands Act. As such, CCROs are rights exercised at the Village Council and Village Assembly level.

"The Hadzabe tried to formalize their own CCROs by annexing from village lands, but for many of the villages, the Hadzabe population was too small and marginalized to get CCROs voted through by the majority non-Hadzabe Village Council and Assembly members," Edward Lekaita, a legal advisor to UCRT, recalls. "But in Domanga and Mongo wa Mono and Yaeda Chini villages, there were enough Hadzabe on the Village Councils and Assemblies to push through a landmark vote for three *communal* CCROs, totaling 34,000 ha and providing the Hadzabe with the rights to live, manage, and use the lands in perpetuity. In the case of Yaeda Chini, the Hadzabe share the CCRO with pastoralists and the CCRO is issued under the name of the village, but the Hadzabe have the rights to practice traditional hunting there. To this day, the Hadzabe are, as far as I know, the only cultural or ethnic group who have been issued CCROs."

Moreover, a CCRO is a stronger and less easily subdivided title than the original village land titles. Selling land under a communal CCRO requires the agreement of the entire group; it is thus highly unlikely for land under a CCRO to be subdivided. This functions to protect the rights of individuals that are reliant on communal land and vulnerable to land grabs, such as women, children, and minorities. This restriction also increases the likelihood the communal land will remain in its current form into the future, as gathering consensus can be challenging. A possible consequence of this restriction, however, is that any community member seeking to divest from the land is unable to do so, and as a result there may be limitations in leveraging the value of the land for access to credit or to use it as collateral for other purposes.

In the Hadzabe CCROs, fencing, charcoaling, and conversion of lands to agriculture or permanent livestock enclosures are explicitly forbidden and regulated under village bylaws. UCRT has supported the establishment of carbon projects in some of the areas, which are designed to foster sustainable land use. "Here, village scouts are employed and paid through the carbon revenues to enforce the bylaws," says Daudi Peterson, a co-founder of Dorobo Tours and author of a book on the Hadzabe. "Traditional Hadzabe woodlands are slowly restoring and, with them, wildlife populations are increasing. Before the project, the Hadzabe lamented the enormous reductions in the wildlife they once hunted. But only a few weeks ago, one of the Hadzabe camps reported hunting an eland, a greater kudu, and a wild pig within one week!" The CCROs appear to be helping to restore the cultural identity of the Hadzabe as hunter-gatherers.

Since the recognition of the original three communal titles in 2011, the Hadzabe have secured the rights to four more communal CCROs for their exclusive use (totaling 2700 ha) and another seven (totaling 33,000 ha) to share with pastoralists (Fig. 12.1). UCRT's work has demonstrated CCROs as an effective route to securing communal tenure claims in Tanzania, and in turn, there is building evidence that communal titles can be an effective pathway to empowering Hadzabe people, protecting their marginalized members, and underpinning the return of biodiversity to the area.

Fragmenting the *Manyatta:* How Privatization and the Subdivision of Lands Blocked Collective Action on the Mara

When Jackson "Jack" Marubu, The Nature Conservancy's (TNC's) project lead for the Maasai Mara, first went to the Maasai Mara, the area was characterized by conflict. "Maasai community members would meet with

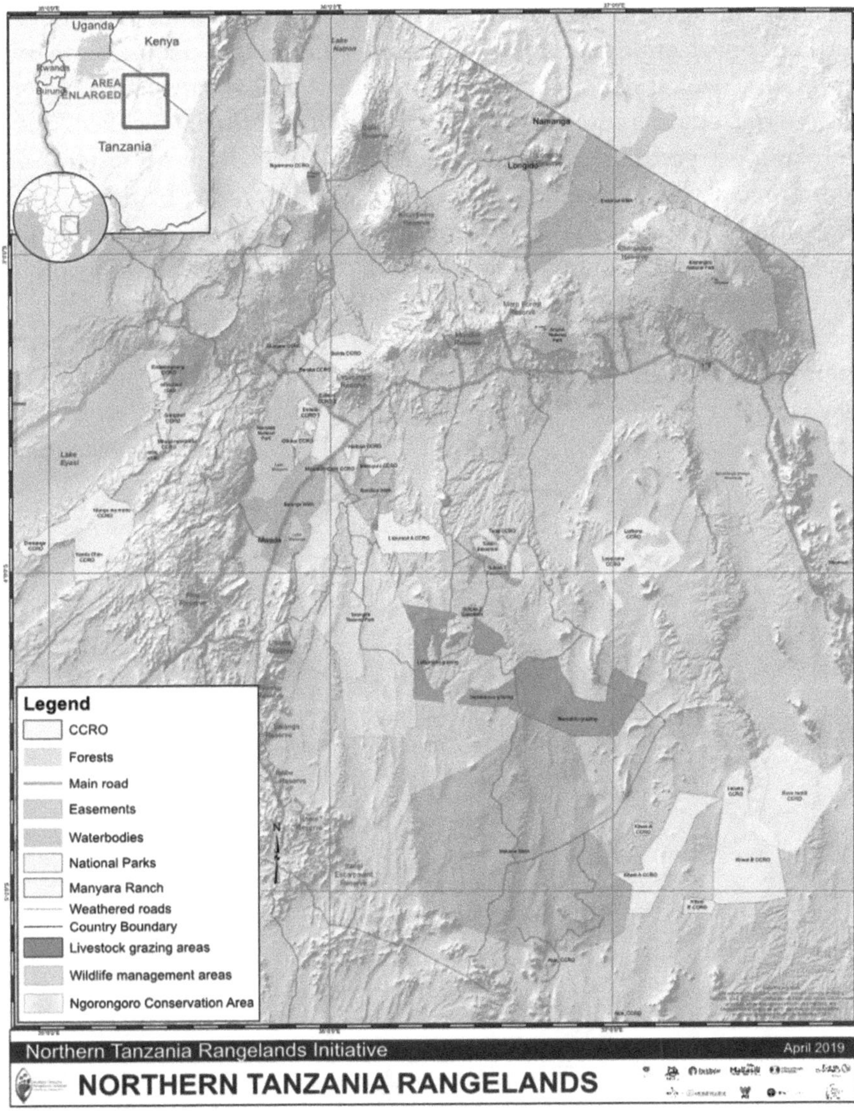

Fig. 12.1 CCROs and national parks across the landscape (NTRI, 2019)

conservation agencies and government authorities and tourism operators, and the conversations would be mayhem—hundreds of people filing into townhalls," Jack says. "Everyone was arguing their separate points of view and no one could listen or agree, mobilization was totally blocked."

Like most pastoralist cultures, Maasai traditional tenure systems are communal, living in mobile villages called *Manyatta*. The British colonial administration perceived Maasai nomadism to be a threat to resource management and sought to develop policies that oppressed pastoralist community movements and communal life, and coerced downsizing to the individual level (Seno & Shaw, 2002). Kenyan Republic administrations inherited these policies and imposed a Group Ranch system on the Maasai Mara in the 1970s to encourage more sedentary lifestyles and commercialize livestock production (Fig. 12.2). This was eventually

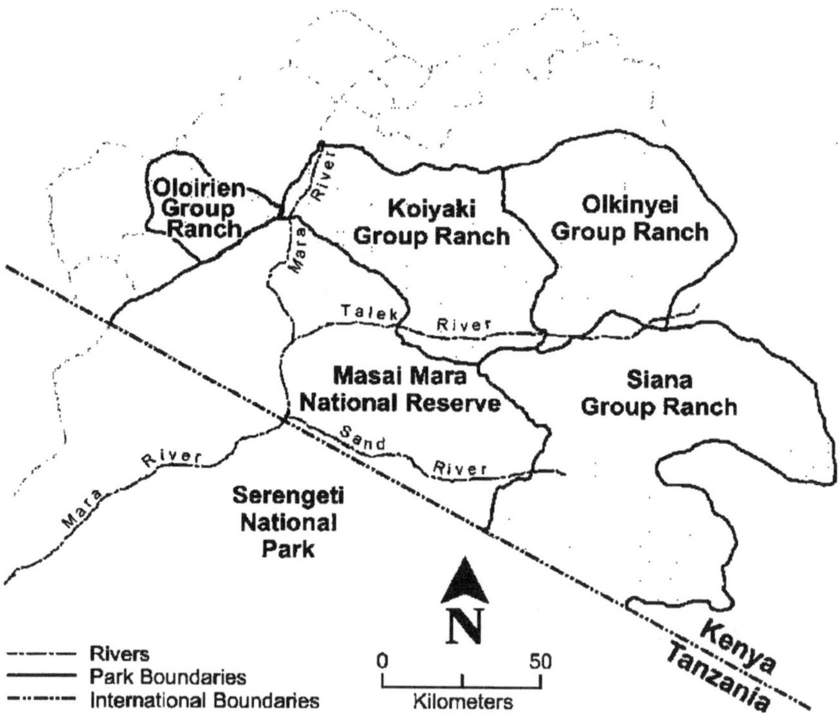

Fig. 12.2 Group ranch system for from the early 2000s. Source: Seno and Shaw (2002)

subdivided further into an Individual Ranch System during the 2000s. This highly privatized system awarded the head of the household, usually a male, the right to sell or lease the land that they claimed.

In 2002, Seno and Shaw conducted a study on Maasai attitudes toward the oncoming privatized subdivision where fencing of individual lands was likely. Although 82% supported it overall as a way to secure land and protect it from encroachment, many supporters (56%) were also concerned privatization would have negative effects, citing reduced resources for livestock grazing and potential limitations to sustaining a livelihood as their main fears. Many were also concerned that dividing lands with fences would have negative consequences for wildlife. Seno and Shaw themselves warned that many Maasai may become landless after selling their lands (4% stated they would sell their tenure rights), predicting that increasing droughts would set in motion a mechanism by which speculators would take advantage of the Maasai's nomadic disinclination to see land as a commodity.

Many of the predictions from this study have become a reality: a mechanism was indeed set in motion whereby supportive government policies promoted the privatization of rangelands, and in response individuals sought to claim lands and pressed for further subdivision to gain individual land titles (Osano et al., 2013).

"Due poor management of lands, the Maasai mistrusted their elected Group Ranch governance and pushed for further subdivision into the individual ranch system. Many of these land parcels were either fenced for land claims or sold to non-Maasai, especially along accessible routes," Jack recalls, "Some conservancies, such as Ol Kinyei, are still threatened by this wave of land sales and subsequent land-use changes."

In parallel to the privatization of surrounding rangelands, the Maasai Mara National Reserve at the heart of the landscape was attracting more and more tourists, currently up to 2700 a day (Broekhuis, 2018), and subsequently increasing revenues. But despite revenues accrued, the surrounding populations remained poor. In 2005, the Central Bureau of Statistics published a poverty study that calculated a 63% poverty rate within 25 km of the park (Central Bureau of Statistics, 2005). Tensions from this inequality were magnified by an exponential population growth, punctuated by the 2008 droughts and further exacerbated by cattle disease that exacerbated declines in wealth.

In an effort to grow grass banks for livestock (Løvschal et al., 2016) and protect them from zoonotic diseases, Maasai and non-Maasai herders erected yet more fences (Fig. 12.3). To access more grass and water, herders encroached into wildlife grasslands further shrinking and fragmenting wildlife habitat. More fences were built by conservation agencies to prevent poaching, illegal resource extraction, and human-wildlife conflict. Ultimately, these fences hindered access to vital resources for livestock and wildlife, caused animal entanglements, and altered species' breeding behavior.

The privatization of Maasai lands (Figs. 12.3 and 12.4) exacerbated by poor, opaque, and corrupt governance fueled further land grabbing and elite capture with powerful individuals amassing extensive lands (Mwangi,

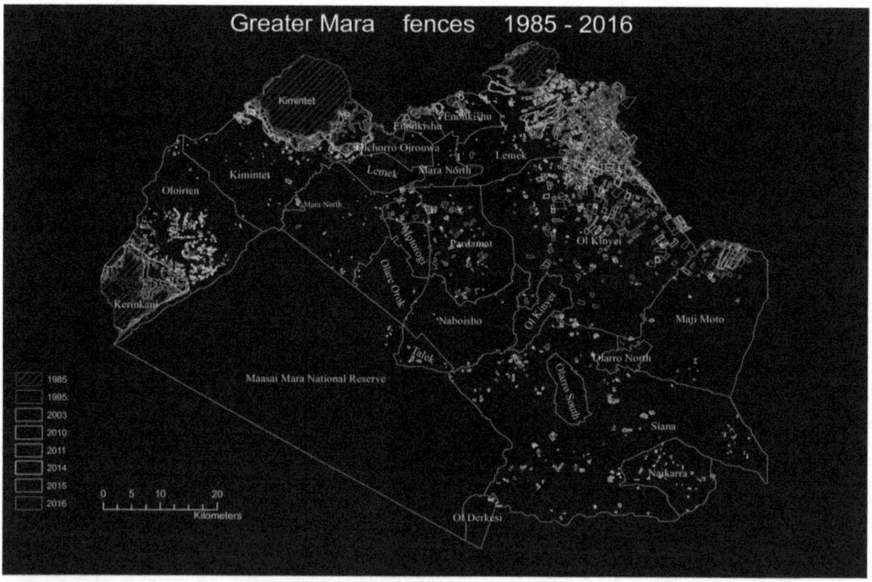

Fig. 12.3 Fences registered on the satellite images (1985–2016). Each year is shown with a distinct color. The year 1985 is marked with a hatched symbol to emphasize the large, densely fenced areas on the periphery. Source: Løvschal et al. (2016)

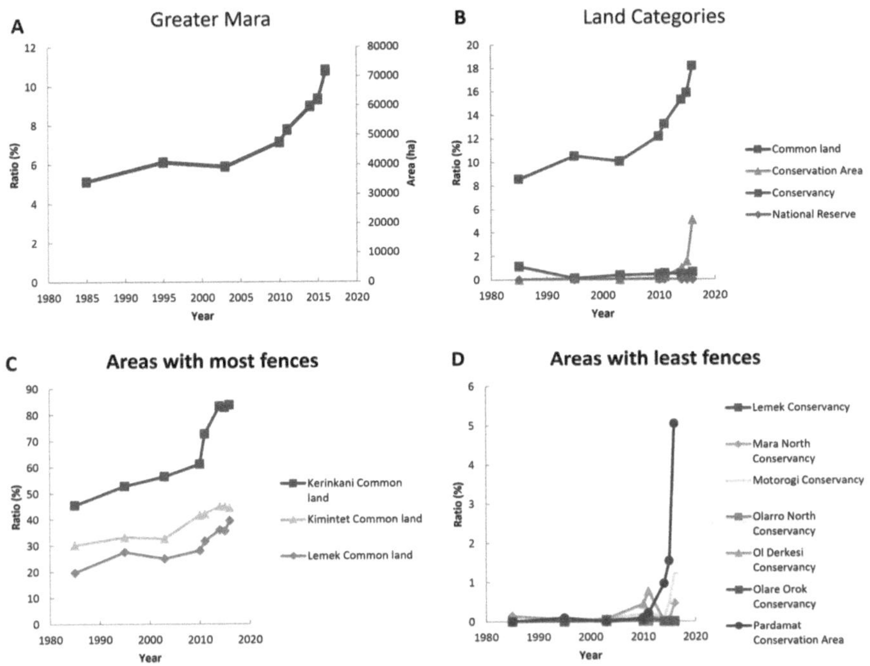

Fig. 12.4 The development of fenced areas for the Greater Mara, as well as for the individual areas. (A) Fenced area of the whole Greater Mara in absolute and relative coverage. (B, C, D) Percent coverage of fences by land category and area. Source: Løvschal et al. (2016)

2007a). These failures compounded in a lack of community cohesion and increased distrust (Mwangi, 2007b).

Figure 12.4 illustrates the rapid increase in fencing, symptomatic of subdivision and privatization, for the Greater Mara ecosystems between 1985 and 2016. Pardamat Conservancy Area (located in the G5 and H5 quadrants in Fig. 12.3) is the most heavily settled and has suffered greatly from an increase in fencing. "During the subdivision of Group Ranches, much of the Mara was uninhabitable due to Tsetse fly and a lack of road networks," remembers Jack, "What is now Pardamat Conservancy Area was deemed the most hospitable area and parcels here were offered to communities for permanent homesteads."

In their study, Løvschal et al. warned of a "critical transition to a chronic landscape shift" characterized by a redefinition of relationships between people and land rendering further subdivision unstoppable and determining boundaries that, once set, do not disappear. If trends did not change, Løvschal et al. predicted a collapse of pastoralism, semi-nomadic lifestyles, and the great migration.

In short, the 140,000 ha iconic landscape of the Maasai Mara was fractured into 14,528 individual parcels and divided by fences. Tenure for many community members was secure, but the resulting division blocked migration routes and the possibility of developing a common vision to manage community lands (Lamprey & Reid, 2004).

Collaborative Civil Society: Maasai Mara Wildlife Conservancies Association Busting down a Siloed Landscape and Breaking Tragedy of the Commons Mythology

In 2001, the Trans Mara County Council established and contracted what is now Mara North Conservancy to manage its portion of the reserve (Walpole & Leader-Williams, 2001). Five years later, the Olare Orok community and Ol Kinyei Group Ranch established two conservancies for a combined 14,576 ha and began working together to broker lease agreements with commercial tourism operators.

> In 2013, a new Wildlife Conservation and Management Act articulated a vision of community conservancies as the instrument for protecting and managing wildlife outside designated protected areas; conservancies were finally given protected area status by law. The Kenya Wildlife Conservancies Association (KWCA) was established in the same year to drive the community conservation agenda at the National level and, with this, a need grew to establish regional bodies to coordinate conservancies and give them a forum to voice their views on conservation. Later that year, MMWCA was established as the regional body for the Mara conservancies, tasked with strengthening conservancy management and governance and uniting conservancy voices.– *Jack Marubu*

Since then, the MMWCA platform has been creating pathways for collective action on the Mara between local communities, landowners, conservancy boards and managers, community institutions, conservation partners, scientific experts, and government institutions including the Kenya Wildlife Service and the Narok County Government (Sopia et al., 2019). In the six years since its establishment, MMWCA has grown to a total of 15 conservancies covering an area of 140,000 ha—a 2280% increase (Fig. 12.5)—with more than 40 tourism facilities paying monthly lease fees directly to the landowners.

MMWCA's wildlife conservancies are unique in that they are made up of around 15,000 individual plots owned by Maasai residents of the Mara and Olderkesi, bringing together thousands of landowners under 15 lease

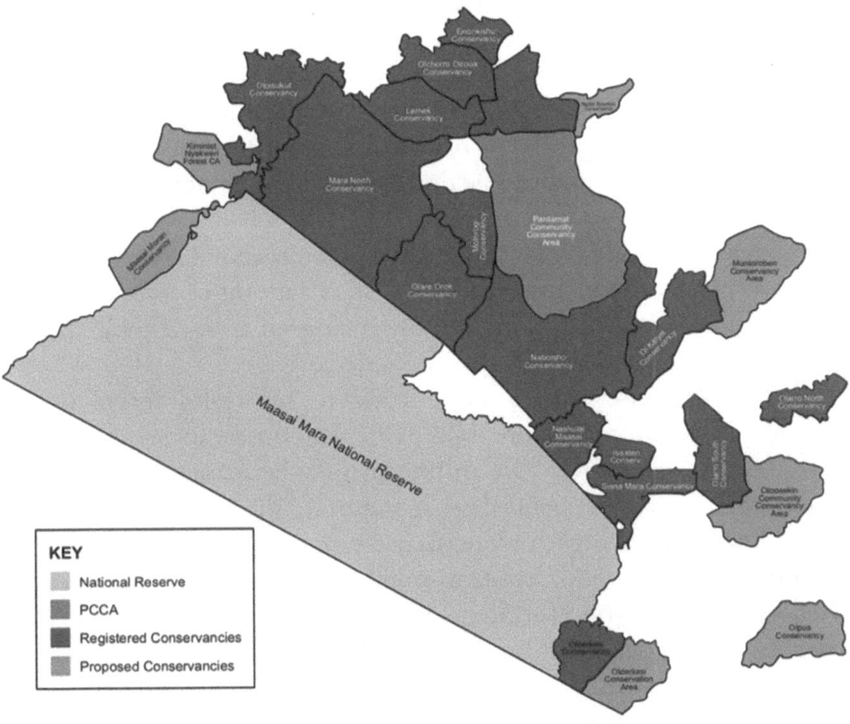

Fig. 12.5 MMWCA map of conservancies and the Maasai Mara National Reserve. Source: MMWCA (2018)

agreements with their respective conservancies. "The leases have clearly defined durations (usually 25 years), easements, and lease payments that go directly to the landowners' bank accounts monthly," says Jack. "The lease payments come from 39 tourism partners and total US$7.5 million annually. And what's more, the agreements limit land sales, protecting not only the current landowners but future generations as well."

Since its inception, MMWCA has also been providing a Livestock to Market program that has developed sustainable grazing plans and empowered herders with leverage to attain a higher price for better-quality livestock products. This is critical for creating strong incentives to participate in the program, one that aligns with their primary livelihood activity. These combined benefits for communities are incentives to participate in actively protecting conservancies. "The expectation is that tolerance for wildlife is boosted and landowners stop activities like fencing, poaching, and any infrastructure projects that will negatively affect wildlife," says Jack. "The conservation agreement is then fully enforced by the conservancies who are managed predominantly by local landowners. Legally, conservancies are also protected from subdivision and alienation."

The increasingly unpredictable droughts due to climate change are also a driver to join the MMWCA. Herders cannot rely on rains to produce grass to feed their cattle and shoats. This phenomenon is forcing herders down one of two paths: joining MMWCA as a means of securing a more reliable income (i.e., communalize) or fencing off areas of their lands to grow grass banks for livestock (i.e., privatize).

MMWCA's conservancies are starting to see real impacts on the ground. "There are now more wildlife in the conservancies than in the Mara Reserve," Jack says. "We think this may be because livestock and wildlife are mutually beneficial." For example, the pastoral practices of "*bomas*," or livestock night corrals, may create nutrient hotspots for wildlife, and removing woody cover to build *bomas* may create habitat preferred by wild herbivores for predator detection (Riginos et al., 2012). In addition, the Løvschal et al. study found that, although fencing in the broader landscape has gotten worse over time, within conservancy areas fencing actually decreased (Fig. 12.4).

Ultimately, the Mara conservancies have a long way to go toward securing ecological goals; as predicted, fences are becoming an increasingly urgent threat in areas surrounding conservancies. However, MMWCA offers avenues for collaborative change and for Maasai to take collective action to secure the future of the Mara. Conservancies provide an ideal tool for community development and conservation. This arrangement is now restoring both Maasai culture and wildlife.

Conclusions and Recommendations: Communal Tenure, Collaborative Platforms, Conservation Incentives, and Explicit Management Agreements

We began our study by asking how different tenurial systems contribute to community-based conservation and aimed to explore two examples to shed light on this mechanism and provide lessons on challenges and opportunities. Our case studies highlight the benefits of communal tenure, as it builds resilience to threats against unsustainable land use that can threaten traditional livelihoods and conservation goals. Our cases highlighted how supporting the Hadzabe to obtain communal titles led to improved human well-being through secured access to resources, as well as forest regrowth and indicative improvements to biodiversity with the return of certain species. Here, communal tenure is providing an environment for the Hadzabe to heal and continue their traditional lifestyles.

We believe the Mara case study is a warning against top-down privatization of lands where it is culturally misaligned. Here, the subdivision of tenurial titles caused a fencing epidemic and redefined the relationship between a traditionally nomadic people and their land, blocking wildlife migration routes and the possibility of developing a common vision to manage the area collectively. But, here again, MMWCA provides an example of how platforms for collaboration can catalyze collective action toward sustainable land uses. In this case, participants saw social fractures diminish, and the MMWCA created avenues for mobilization to reduce fencing and spur sustainable management of the area.

A key assumption in these types of programs is that incentives—whether traditional, economic, or both—are aligned with sustainably using the land at the community level. The Hadzabe utilized the democratic governance structure of the CCROs to exercise greater control over their lands to support traditional resource use and thus an ecologically lighter use of the area. The MMWCA platform bolstered collaboration among the Maasai communities, allowing the Maasai to feel secure in their access to land in the future and reducing the atmosphere of land grabbing and subsequent fencing.

The legal mechanisms for titling to support community-based activities in our case studies highlight how legal frameworks can serve as a critical element to responsible management plans for managing resources sustainably. Hadzabe CCRO bylaws explicitly forbid fencing, charcoaling, and conversion of lands to agriculture or permanent livestock enclosures, while the Maasai participate in actively protecting conservancies and stop activities like fencing, poaching, and any infrastructure projects that will negatively affect wildlife. Laws and regulations do not always align with community-based management, but when they do and are widely recognized and consistently enforced, the case studies demonstrate how they can serve as a strong mechanism to support sustainable resource use through increasing tenure security.

While legal titles do not work as a direct incentive in either of our case studies, titles provided a platform to organize incentives. Hadzabe CCROs are supporting forest carbon projects from which the Hadzabe collect revenues, stopping the felling of 12,000 trees annually. Meanwhile, MMWCA's lease agreements provide financial incentives to landholders from tourism.

Considering the above findings, we recommend community-based conservation projects seeking to attain human well-being and conservation goals through securing tenure complement this approach with:

- conservation incentives linked to tenurial parcels;
- explicit sustainable management agreements ideally linked to the conservation incentives;
- a collaborative platform to support community mobilization and project participation; and
- a sustainable national financing mechanism.

We also recommend projects that support tenurial systems that are sensitive to, and in keeping with, communities' traditions and cultures. As with our case studies, highly privatized tenure systems may not be appropriate to all communities and may undermine project goals.

References

Adams, W. M., & Hulme, D. (2001). If community conservation is the answer in Africa, what is the question? *Oryx, 1*, 193–200.

Barrow, Edmund, and Marshal Murphree. 2001. "Community conservation: From concept to practice." In African wildlife and livelihoods: The promise and performance of community conservation, by D. Hulme and M. Murphree, 24–37. Heinemann.

BirdLife International. (2019). Important Bird Areas factsheet: Masai Mara. Retrieved December 10, 2019, from http://datazone.birdlife.org/site/factsheet/masai-mara-iba-kenya.

Breckheimer, I., Haddad, N. M., Morris, W. F., Trainor, A. M., Fields, W. R., Jobe, T. R., Hudgens, B. R., Moody, A., & Walters, J. R. (2014). Defining and evaluating the umbrella species concept for conserving and restoring landscape connectivity. *Conservation Biology, 1*, 1584–1593.

Broekhuis, F. (2018). Natural and anthropogenic drivers of cub recruitment in a large carnivore. *Ecology and Evolution, 1*, 6748–6755.

Brooks, J., Waylen, A., & Mulder, M. B. (2013). Assessing community-based conservation projects: A systematic review and multilevel analysis of attitudinal, behavioral, ecological, and economic outcomes. *Environmental Evidence.*

Central Bureau Of Statistics. (2005). *Geographic dimensions of well-being in Kenya: Who and where are the poor? A constituency level profile. Vol. II.* Central Bureau of Statistics, Ministry of Planning and National Development.

KWCA. (2020). *Kenya Wildlife Conservancies Association.* Retrieved October 02, 2020, from https://kwcakenya.com/conservancies/.

Lamprey, R. H., & Reid, R. S. (2004). Expansion of human settlement in Kenya's Maasai Mara: What future for pastoralism and wildlife? *Journal of Biogeography, 1*, 997–1032.

Løvschal, Mette, Bøcher, Peder Klith, Pilgaard, Jeppe, Amoke, Irene, Odingo, Alice, Thuo, Aggrey, & Svenning, Jens-Christian. (2016). Fencing bodes a rapid collapse of the unique Greater Mara ecosystem. *Nature: Scientific Reports.*

MMWCA. (2018). *2018 Impact Report.* Annual Report, Narok: Maasai Mara Wildlife Conservancies.

Mwangi, E. (2007a). Subdividing the commons: Distributional conflict in the transition from collective to individual property rights in Kenya's Maasailand. *World Development, 1*, 815–834.

Mwangi, E. (2007b). The puzzle of group ranch subdivision in Kenya's Maasailand. *Development and Change, 1*, 889–910.

Northern Tanzania Rangelands Initiative. (2019). *Rangelands in transition: Sustaining healthy landscapes for people and wildlife in northern Tanzania.* NTRI.

Northern Tanzania Rangelands Initiative. (n.d.). *A New Way for People and Wildlife to Thrive.* Narrative donor report, Arusha: USAID.

Osano, P. M., Said, M. Y., de Leeuw, J., Ndiwa, N., Kaelo, D., Schomers, S., Birner, R., & Ogutu, J. O. (2013). Why keep lions instead of livestock? Assessing wildlife tourism-based payment for ecosystem services involving herders in the Maasai Mara, Kenya. *Natural Resources Forum, 1*, 242–256.

Redfern, K. (2018). Securing Hadza land titles, securing futures in Tanzania. *Cultural Survival Quarterly Magazine.*

Riginos, Corinna, Porensky, Lauren M., Veblen, Kari E., Odadi, Wilfred O., Sensenig, Ryan L., Kimuyu, Duncan, Keesing, Felicia, Wilkerson, Marit L., & Young, Truman P.. (2012). Lessons on the relationship between livestock husbandry and biodiversity from the Kenya Long-term Exclosure Experiment (KLEE). *Pastoralism: Research, Policy and Practice.*

Robinson, B. E., Masuda, Y. J., Kelly, A., Holland, M. B., Bedford, C., Childress, M., Fletschner, D., et al. (2017). Incorporating land tenure security into conservation. *Conservation Letters, 1*, 1–12.

Schuette, P., Creel, S., & Christianson, D. (2013). Coexistence of African lions, livestock, and people in a landscape with variable human land use and seasonal movements. *Biological Conservation, 12*, 148–154.

Seno, S. K., & Shaw, W. W. (2002). Land tenure policies, Maasai traditions, and wildlife conservation in Kenya. *Society & Natural Resources, 1*, 79–88.

Sopia, Daniel, Reson, Eric Ole, & Mwangangi, Jones. (2019). *State of Mara Conservancies Report.* Annual, Narok: Maasai Mara Wildlife Conservancies Association (MMWCA).

Ujamaa Community Resource Team. (2014). *Securing communal land tenure in northern Tanzania using certificates of customary right of occupancy.* UCRT.

Walpole, M. J., & Leader-Williams, N. (2001). Masai Mara tourism reveals partnership benefits. *Nature, 771*, 1.

Western, D., Russell, S., Cuthill, I., & Somers, M. (2009). The status of wildlife in protected areas compared to non-protected areas of Kenya (wildlife trends across Kenya). *PLoS One, 1*, 6140.

Williams, A. (2017). *Community natural resource Management in Tanzania.* Bureau for Africa/Office of Sustainable Development: USAID.

13

Beyond the Traditional: Voluntary Market-Based Initiatives to Promote Land Tenure Security

Daniela A. Miteva, Lea Fortmann, and Roan McNab

Tropical developing countries are important for the provision of global ecosystem services like climate change mitigation, support for biodiversity, and the regulation of global hydrological flows; however, they are also experiencing high poverty levels and rapid destruction of the natural resources on which they depend (MEA, 2005). In many developing countries, formal institutions are weak or non-existent. Informal institutions may be insufficient as well depending on the context (e.g., large scales, heterogeneous populations, migration, and displacement due to military conflicts) (Baland & Platteau, 1996). A promising new approach to address tenure insecurity in tropical countries in the absence of strong

D. A. Miteva (✉)
The Ohio State University, Columbus, OH, USA
e-mail: miteva.2@osu.edu

L. Fortmann
University of Puget Sound, Tacoma, WA, USA

R. McNab
Wildlife Conservation Society, Guatemala City, Guatemala

M. B. Holland et al. (eds.), *Land Tenure Security and Sustainable Development*,
https://doi.org/10.1007/978-3-030-81881-4_13

institutions is the voluntary market-based interventions that have emerged in the past 30 years (e.g., Auld et al., 2008; Milder et al., 2015). These are a class of interventions like commodity certification that use markets to provide incentives for participation like price premiums or market access for primary commodity producers and/or supply chains. Their take-up is voluntary, but once enrolled in the certification scheme, producers have to comply with a set of standards. Many of these certification schemes include provisions about resolving land tenure conflicts.

The goal of this chapter is to review the role of voluntary market-based interventions as a mechanism for ameliorating forest land tenure insecurity in the absence of strong local institutions in developing tropical countries. We specifically focus on one such market-based intervention—the Forest Stewardship Council (FSC) forest management certification as it directly affects working forest lands, has been around for nearly 30 years, and is the most common such intervention in tropical countries (Auld et al., 2008; Potts et al., 2014); as of late 2017, 198 million hectares of forests are under FSC certification globally, with 16% located in Asia, Africa, and Latin America (source: Mongabay: https://news.mongabay.com/2017/09/does-forest-certification-really-work/ accessed January 11, 2020). However, our insights can be applied to other similar voluntary market-based interventions. We focus exclusively on native forests as they are of global conservation importance and are most affected by tenure insecurity relative to other forest land uses like agroforestry and tree plantations.[1]

The Causes of Forest Land Tenure Insecurities in Developing Countries

Even though forest lands are of primary importance for supporting livelihoods, providing key ecosystem services, and generating revenue in most tropical developing countries (MEA, 2005), forest land management is

---

[1] For example, forest plantations may be less affected by tenure insecurity because tree planting is often seen as laying claims to the land (Fenske, 2011). Further, plantation tree planting may also signal greater tenure security as forest managers do not have incentives to plant trees, if there is a risk that they are going to be uprooted (Fenske, 2011).

often plagued by weak institutions that are unable to clearly define, delineate, and enforce property rights, and punish perpetrators. In this chapter we focus on two types of institutional failures that pertain to working forest lands in tropical developing countries:

1. *Unclear delineation of land rights*, resulting in either confusion as to where one's land is or creating overlapping land rights, with multiple actors claiming use or ownership rights over the same piece of land. These are often caused by land reform policies implemented when inadequate institutional structures exist, in the presence of rapid institutional change, or a combination of both. For example, the rapid decentralization in Indonesia led to the formal recognition of customary land tenure (Larson & Soto, 2008; Resosudarmo, 2004). At the same time, district governments also issued permits to logging companies to raise revenues, often over the same productive forest lands (e.g., Engel et al., 2006; Miteva, 2013; Resosudarmo, 2004), leading to conflict over tenure rights. How these conflicts were resolved depended on the bargaining power of local communities, which in turn depended on the community characteristics and the value of the forest to local communities (Engel et al., 2006; Larson & Soto, 2008; Resosudarmo, 2004). In other locations, like postwar northern Uganda (e.g., see Chap. 8 in this volume), the prolonged military conflict, coupled with the prolonged displacement of a large fraction of the rural population into internally displaced persons (IDP) camps, resulted in the breakdown of the previous informal institutions, with no adequate formal institutions to replace them (Bjørkhaug et al., 2007; Deininger & Castagnini, 2006; Miteva & Brown, 2018; Miteva et al., 2019). Unsurprisingly, postwar, there has been significant conflict over land, exacerbated by a land tenure reform encouraging the conversion of the traditional customary land ownership, regulated by clans, to private land ownership, with a household holding the deeds to land that can be traded, transferred, or used as collateral for loans (Bjørkhaug et al., 2007; Deininger & Castagnini, 2006; World Bank, 2009). Because of the lack of formal and informal institutions in postwar northern Uganda, there have been a lot of instances of land grabbing, especially by wealthy male-headed households (World Bank, 2009; Bjørkhaug

et al., 2007; Adelman & Peterman, 2014; Miteva & Brown, 2018). Clearing of native forests and their subsequent conversion to agriculture is a way to establish claims over the land (Fenske, 2011; Unruh et al., 2005; Lambin et al., 2018; Miteva et al., 2019).

2. *The inability to enforce existing regulations*. For example, in areas with clearly defined property rights in Indonesia, households often use rivers to transport illegally harvested trees from logging concessions or protected areas (Resosudarmo et al., 2012). It has been estimated that illegal logging contributed to 64% and 83% of the total timber production in 2000 and 2001, respectively (Resosudarmo et al., 2012). Similarly, even though land property rights are clearly defined and delineated in Mexico, there is illegal logging in community-managed forest lands (e.g., ejidos in Mexico) (e.g., Alix-Garcia et al., 2005; Honey-Rosés 2009). Even if done selectively for only the most valuable species, it may still hamper efforts for the long-term management of the land as forest by reducing the incentives to keep the forest intact. How the ejido members address illegal logging depends on the characteristics of the community, the value of timber, and the ease of monitoring (e.g., Alix-Garcia, 2005; Baland & Platteau, 1996). While some ejidos may address the issue by increasing the monitoring of forests or bribing non-ejido members to protect forests (e.g., Alix-Garcia et al., 2005), others may resort to the parceling of land and the creation of individual land rights (Ellis et al., 2015). Because commercial forestry requires large scales, the parceling of land may lead to deforestation since commercial forestry may not be profitable at a small scale.

In both cases, the insecure land rights over forests often result in (a) increased deforestation as a way to establish claims to the land (e.g., in Uganda) or minimize losses from illegal logging (e.g., Mexico) and (b) decreased incentives to invest in the land by replanting trees or improving the soil quality. That is, land tenure insecurity in tropical developing countries often leads to forest degradation and deforestation. The goal of this chapter is to illustrate how a voluntary market-based initiative like forest management certification can improve land tenure security and improve forest conservation on the ground.

FSC Certification as a Way to Address Forest Land Tenure Insecurity and Promote Forest Conservation

In this chapter, we argue that FSC certification can be a substitute for weak institutions *under certain conditions*. Initiated in the early 1990s by conservation non-governmental organizations (NGOs) in response to rapid deforestation, FSC aims to promote forestry practices that are economically viable, socially beneficial, and environmentally friendly, and thus, contribute to the sustainable management of production forests (Auld et al., 2008).[2] It includes measures to minimize the impact from logging on forests (e.g., restrictions related to the methods and amounts of timber harvesting, road width as well as measures to conserve soils and high conservation value habitat for species), measures to improve worker safety and rights and community well-being, and measures to ensure compliance with laws, recognition of Indigenous People's rights, and land tenure conflict resolution (FSC principles and criteria, 2015). The potential benefits for timber producers and timber processing companies from certification include (a) increased market access, for example, to Europe and the US, which do not allow for non-certified timber to be imported; (b) price premiums; and (c) brand loyalty and increased demand for products (Auld et al., 2008; Breukink et al., 2015).

While the certification is guided by universal principles underlying the three goals, the forestry practices and criteria for each country are developed in accordance to local characteristics (FSC, 2009). Prior to certification and each year post-certification, the FSC forest management units are audited by independent third-party auditors. If the auditors find violations, they issue corrective action requests (CARs); if the prescriptions in those are not met within a specified time period, the FSC certification can be suspended or revoked.

FSC can play a vital role in resolving land tenure insecurities, when the local formal and informal institutions are inadequate. Specifically,

[2] A map of the distribution of FSC certification at a country level can be found here: https://fsc-int.maps.arcgis.com/apps/webappviewer/index.html?id=06188ad39e5344db96a4a181e135c393&mobileBreakPoint=300.

certification requires compliance with the law and international agreements, tenure security, and conflict resolution among all stakeholders, recognition of Indigenous People's rights, and community relations and workers' rights (FSC principles and criteria, 2015). For example, prior to the certification of PT Erna Djuliawati in Indonesia, the auditors selected 11 villages for field visits and interviews with community members; the villages were selected based on the auditors' perceptions where both positive and negative impacts of certification are likely to occur (Smartwood, 2005). If the meetings with local stakeholders indicate unresolved tenure disputes, the concession holders are issued CARs to be resolved prior to certification. Similarly, if new land tenure conflicts arise post-certification, the concession holders are also issued CARs during the annual audits and given a time frame to resolve them. For example, even though PT Sumalindo Lestari, an FSC-certified concession in Indonesia, had a good relationship with the local communities at the time of certification, the influx of new households forming new settlements and practicing slash-and-burn agriculture inside the concession necessitated the issuance of a CAR involving more community building relations and the development of a better system for monitoring illegal logging (Smartwood, 2007). The company complied within a year.

The implementation on the ground often includes a formal benefit-sharing mechanism with local communities (e.g., Cerutti et al., 2017; Vermeulen & Karsenty, 2017). It can be based on just the presence of customary lands within the concession or proportional to their area (Cerutti et al., 2017). Alternatively, concession holders may provide local community development or health programs (e.g., Miteva et al., 2015).

FSC certification is often viewed as a non-state voluntary market-based third-party governance system (Cerutti et al., 2017). The examples above illustrate how FSC certification can be used as a mechanism for conflict resolution and frequent forest monitoring that does not depend on weak governments or ineffective informal institutions. Thus, by addressing tenure insecurity issues, FSC can contribute to the sustainable use and protection of forests. In addition, by providing access to markets or by providing price premiums for certified timber, FSC can also make commercial forestry profitable (e.g., Breukink et al., 2015) and provide a sustained source of local livelihoods. Thus, it can create incentives for land to be kept as forest. For example, in the Maya Biosphere Reserve (MBR) (Guatemala), FSC-certified forestry operations in community-held and industrial concessions generate income that exceeds from agriculture and provides employment for ~25% of the population associated with the concession; in those areas, concession management was also effective in reducing deforestation (Bocci et al., 2018; Fortmann et al., 2017).

Is FSC the Panacea?

We argue that, by creating incentives for the long-term management of forests, FSC *can* be an effective mechanism to promote tenure security and protect forests in tropical developing countries. However, several caveats are in order:

First, FSC targets only production forests; these are lands with valuable timber and/or non-timber forest products (e.g., *Chamaedorea* spp. palm fronds that are exported to Europe and the US), usually at a larger scale, so that commercial forestry is profitable. FSC is not designed to be used in locations that are not commercially viable. Further, certified concessions should have sufficient market access, so that the certified timber and other products can meet the demand. If market access is difficult, the demand for certified timber is low, or consumers find it difficult to distinguish between certified and non-certified timber, FSC is unlikely to provide incentives for the sustainable use of production forests.

Second, while FSC is a global voluntary market-based intervention, with the requirements adjusted to a local context, its implementation often is done on a concession-by-concession basis.[3] That is, even though FSC may provide incentives for a concession holder to resolve the land tenure conflicts around the certified forest tract, the impacts are at a relatively small scale. Unless extensive tracts of productive forest land are placed under FSC certification, large-scale improvements in tenure security due to the intervention are unlikely.

[3] There are some exceptions like the multi-concession certification under the aegis of Forescom, with some 3–4 concessions in the Maya Biosphere Reserve in Guatemala, for example.

Third, FSC tends to be located in areas where complying with the guiding principles and criteria is feasible and compliance is less costly. Forest management units, whose location makes it difficult to monitor and enforce logging restrictions, may choose not to pursue FSC certification as compliance may be very difficult or costly. For example, in Indonesia, relative to traditional logging concessions, the FSC certified ones are placed in isolated areas with lower population density, shorter river networks, which traditionally facilitate illegal logging as the logs are transported down large rivers undetected, and less peatland, which is protected by national law (Miteva et al., 2015). Similarly, areas with high conservation value and valuable timber, but high incidence of large-scale conflicts, like the Democratic Republic of Congo, currently lack any active FSC certification; all previous FSC certificates in the country have been terminated or revoked (FSC, 2019: https://info.fsc.org/certificate.php#result).

Fourth, since FSC certification is voluntary, concession holders may decide not to pursue it if certification involves changing practices for which they have little or no experience and the financial means to implement. Similarly, when compliance with certification becomes costly, concession holders may opt out of certification or have their certification canceled or suspended. The reasons for that include changing market incentives or exogenous change (e.g., influx of migrants or natural disasters like hurricanes wiping out entire forests). In addition, local governments may also limit the effectiveness of forest management certification, for example, by demanding exorbitant bribes (Breukink et al., 2015) or threatening to terminate the concession.

Fifth, certification may not be easily accessible. For example, as certification often necessitates adoption of reduced impact logging and improved worker safety measures as well as certification and auditing costs, certification can be expensive (e.g., Breukink et al., 2015). That is, large commercial concessions holders who have access to large funds, or have already adopted better practices and have resolved tenure conflicts are more likely to get certified; for smallholders and communities, the costs of certification may be prohibitively high (e.g., Lambin et al., 2018).

The caveats described above suggest that the effectiveness of FSC certification as a mechanism to improve tenure security and promote the sustainable use of forests is contingent on the uptake and continuation of certification, which in turn depend on a number of conditions, most notably the costs of certification and compliance and the demand for certified products.

NGOs to the Rescue?

Certification by itself is often not sufficient and requires partnerships with non-governmental organizations (NGOs), in order to effect change on the ground. The goal of this section is to highlight the channels through which NGOs can help address some of the limitations that have been shown to hamper the effectiveness of FSC. We argue that, while not perfect, NGOs can help lower the costs of certification and compliance and increase the demand for certified products, thereby addressing most of the caveats listed above that undermine FSC's effectiveness.

1. NGOs can lower the costs of certification and compliance. For example, since certification often requires substantial changes in the forestry practices, the Tropical Timber Foundation works with concession holders in Indonesia to train staff in terms of reduced impact logging practices and sustainable forestry, before the concession receives certification (source: https://www.tff-indonesia.org/index.php/certification/certification-support2; accessed March 27, 2019). Similarly, before certification in the Maya Biosphere Reserve in Guatemala, a number of communities with relatively small forest management units partnered with NGOs, who provided technical, administrative, and financial assistance (see Box 13.1). In Peru, the Maderacre concession, composed of indigenous Iñapari district inhabitants, obtained FSC certification with the help of the World Wildlife Fund (WWF) and USAID (source: http://www.wwf.org.pe/index.cfm?uNewsID=192806&uLangID=1; accessed March 27, 2019). Further, the concession was able to obtain Reduced Emissions from Avoided Deforestation and Forest Degradation (REDD) funds to offset some of the costs of certification and further supplement the forest area within the concession (Jose Canchaya, 2015, personal communication). Thus, in addition to companies, NGOs can increase the number of smallholders and community concessions that receive certification and remain cer-

tified; for these the costs of certification and compliance might otherwise be prohibitively high.

2. NGOs can facilitate land tenure conflict resolution. For example, driven by NGOs, some FSC-certified concessions in Central Africa have started mapping customary lands spanned by the concession areas (Karsenty & Hardin, 2017). The companies there use this information for revenue sharing with local communities. Similarly, in order to secure FSC certification in the Congo Basin, Congolaise Industrielle des Bois partnered with the Tropical Forest Trust, an NGO that provided technical guidance and assisted with building relationships with the local semi-nomadic peoples whose land the concession overlapped (Watson, 2009). A key aspect of the partnership involved participatory community mapping with Geospatial Positioning System (GPS) units provided to local communities. The latter would walk through the forest and provided spatially explicit maps of natural springs, sacred sites, and areas for harvesting key non-timber forest products. These maps were used to avoid timber harvesting in areas important to the Indigenous Peoples and minimize conflict. Further, the partnership established a local radio station, whose broadcasts are controlled by local people as a way to increase their influence of how the forest is managed (Watson, 2009).
3. NGOs can assist with forest monitoring and enforcement of regulations. Because many NGOs have offices in tropical locations and carry out fieldwork on the ground or have launched auditing programs, they are often instrumental in detecting non-compliance in certified forest management units. For example, even though PT Intracawood obtained FSC certification in 2006, it had its certification suspended multiple times due to non-compliance reported by the auditors, the Rainforest Alliance Smart Wood Program (Smartwood, 2006, 2008). The concession is currently listed as being FSC certified (FSC, 2019). NGOs can also support and/or lead ecological monitoring of certified forests, helping to evaluate the ecological impact and provide recommendations for improved management. For example, the Wildlife Conservation Society and partners evaluated certified forests in Guatemala, Nicaragua, French Guiana, and Bolivia, concluding that at these specific sites certified forest management has been effective in

conserving jaguars and their habitats (Polisar et al., 2016). Studies of this nature can assist in making the case for the continuance of forest management regimes, especially in the face of relentless pressure from alternative land uses including cattle grazing and oil palm cultivation.

4. NGOs can help increase the demand for certified timber. For example, with the help of large consumer awareness campaigns, large international conservation NGOs can exert considerable pressure on manufacturers dependent on timber and, hence, their supply chains (Lambin et al., 2014). They can also help consumers directly differentiate between certified and other timber by lending their logos on the products made from certified timber. For example, in the US WWF's or the Rainforests Alliance's logo can be seen on a host of products from printer and toilet paper to coffee. The placement of the logo of major conservation NGOs signals to consumers that the product is made from certified timber, reducing the need for consumers to research products in depth. Further, NGOs are often instrumental in helping communicate the impacts of certification. For example, recently the ISEAL Alliance, an umbrella organization of sustainability standard holders (https://www.isealalliance.org/about-iseal/who-we-are), formed the Value Added Impact (VIA) initiative with representatives of major NGOs involved in sustainable timber certification as well as academics. The goal of the initiative is for companies purchasing sustainable timber and representatives of the FSC impact and evaluation body to provide guidance about the pertinent policy-relevant questions to businesses and FSC communities. The initiative reviewed the existing literature on the impacts of FSC certification, distilled messaging about the impacts of the intervention based on the existing evidence, and, more importantly, provided guidance about how existing academic studies can be used to inform credible messages about the impact of certification. The latter includes guidance on what constitutes causal impacts versus correlations, what are some potential drawbacks of existing studies, as well as the appropriate vocabulary to use to describe different types of studies in business communications. The outputs from the VIA initiative are intended to help businesses communicate *in a credible way* the impact of certification to consumers, and, thus, to increase the demand for certified products and the uptake of certification.

Box 13.1 Case Study: Community Forest Concessions of Guatemala's Maya Biosphere Reserve

One example of a hybrid partnership is the case of the community forest concessions in Guatemala, where the local government requires that the communities become FSC certified in order to maintain their concession status. Two decades of experience with the development and management of the community forest concessions in Guatemala's Maya Biosphere Reserve have highlighted some of the benefits of certification as a mechanism for improved land tenure security discussed in this chapter, as well as its challenges and limitations. Both international and local NGOs played a crucial role in providing technical and financial support to the concessions as they developed. NGO partners continue to offer support, with their roles evolving over time as local concession managers have gained experience and become more established, and as the second-tier organization known as the "Association of Forest Communities of Peten" or ACOFOP plays an increasingly important role in technical assistance (Fig. 13.1).

While the MBR was established in 1990, the first community forest concession was not created until 1994. The impetus for the concessions was to provide local populations access to forest resources; however, diverse actors were concerned about the potential ecological impacts, the socioeconomic

Fig. 13.1 FSC-certified timber in the Maya Biosphere Reserve. (Credit: Ben Schilling/WCS)

viability of forest management, the ostensible lack of financial capital faced by local communities, and by the long-term financial sustainability of forest management. For these reasons, a 1992 study funded by USAID and developed by Tropical Research and Development concluded that local communities would not be able to manage vast forest tracts, recommending instead that the communities be provided areas of 5000 hectares maximum, to use as wood lots and for ensuring their subsistence needs.

After a pilot program led by the Center for Tropical Agriculture and Education (CATIE: https://www.catie.ac.cr) established the first community forest concession of 7039 hectares in San Miguel la Palotada, the tide began to turn and public and professional sentiment in Guatemala began tilting toward greater community control. The coupling of community concessions with FSC certification ended up being a key factor in the Guatemalan government's final decision to prioritize community concessions under a hybrid partnership in which CONAP, the local governance institution, allowed the concessions to form under the condition that they become FSC certified within the first three years of formation (Monterroso & Barry, 2012). Communities that were awarded forest concessions were allocated property rights to the land for a renewable period of 25 years as long as they maintained their certification status (Fig. 13.2). Previously, local communities had no formal *de facto* rights in the area whatsoever; but once concessions were granted communities had legal recourse to resist, and eventually withstand considerable pressure brought by large-scale tourism and their interest groups that, as of 2002, began to promote a large archeological "wilderness area" entitled the "Mirador Basin" and advocate for strict conservation in the region (Nittler & Tschinkel, 2005; Radachowsky et al., 2012).

Given the lack of sustainable forest management experience among many of the local communities, one of the requirements for gaining concession status was that the community partner with an NGO that would provide technical, financial, and administrative assistance (Radachowsky et al., 2012). One of the first steps to gaining concession status was for the community to legally register as an association. With the help of their partner NGO, the community concession members were guided through this process, along with developing an environmental management plan for the forest area, which is required for FSC certification. In addition to lack of technical experience, many of the concession managers also lacked business acumen. The NGOs were able to provide additional support and expertise when it came to matters such as financial planning, paying taxes, meeting labor standards, writing labor contracts, and registering workers for social security.

As noted earlier in the chapter, a number of caveats exist that determine whether certification is a good fit for the forest management unit and whether it will be successful in providing environmental protection and economic support for the local communities. These caveats apply to the experience of the concessions in the MBR. While 12 concessions were ini-

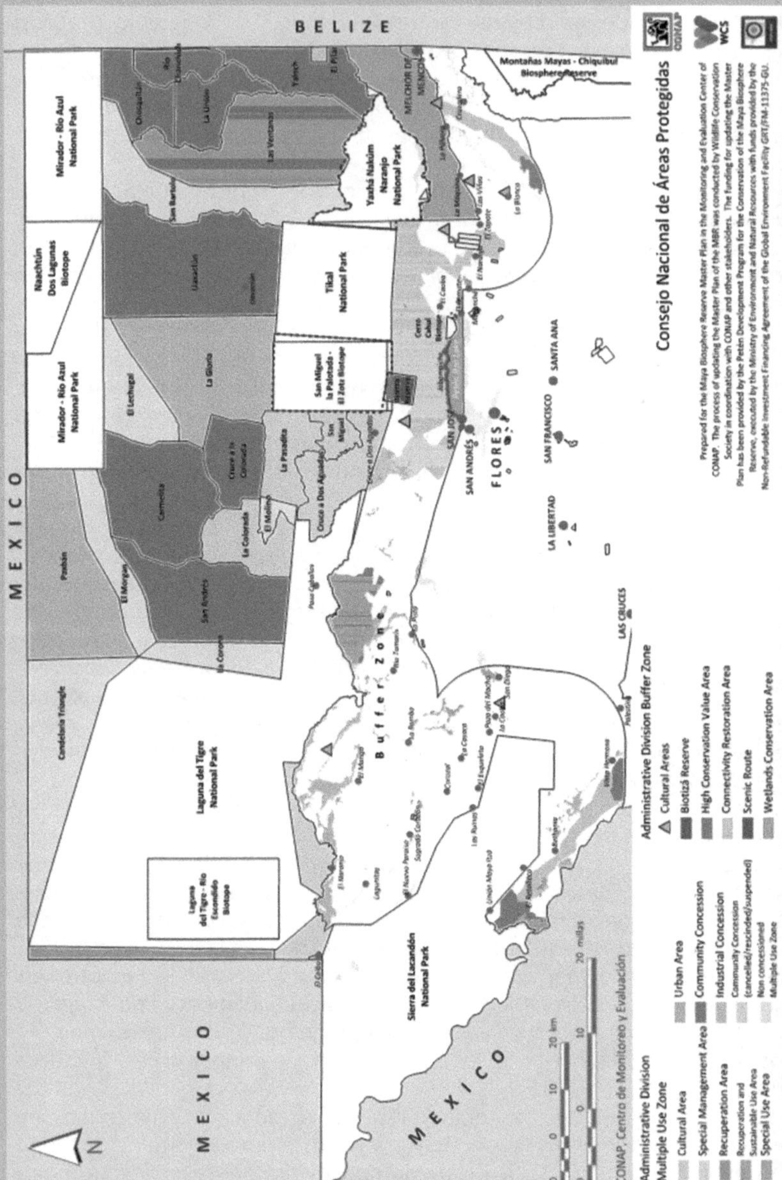

Fig. 13.2 Location of community-based and industrial concessions (with FSC certification, all 11) within the Maya Biosphere Reserve

tially formed, 3 have since been cancelled or suspended, and one requires consistent technical and financial support to continue due to land conflicts and considerable deforestation in the area. A number of factors contributed to the success of some of the concessions and the fact that others have struggled from the beginning. The concessions that have been most successful tended to be the nonresident concessions composed of members living outside the concession boundaries (Radachowsky et al., 2012). These groups came together voluntarily to form a concession out of common interest in forest management, and many of the members had previous forestry experience. Additionally, the more successful groups also had stronger local and social ties, including two concession communities that have been long established in the Petén and have historically depended on the forests and its resources for their livelihoods (Gómez & Méndez, 2007; Fortmann et al., 2017).

The concessions that have failed are primarily the recently inhabited concessions made up of members coming from predominantly agricultural backgrounds. These members were initially not interested in forest management, but ultimately came together to form a concession out of desire to maintain their landholdings within the reserve and at the prompting of their partner NGOs (Nittler & Tschinkel, 2005). Another factor that contributed to the downfall of these concessions was their relatively smaller size and paucity of high-value timber, which would make it harder to extract enough to be economically sustainable and cover the cost of compliance to maintain their certification. To try to overcome some of these issues, a group of concessions joined together to be certified as one forest management unit to spread out the costs, but this also meant that the success or failure of each concession was tied to the success or failure of the group as a whole. A number of external factors also played into the breakdown of the recently inhabited concessions. These areas were under greater pressure from external forces, including cattle ranchers and narco drug traffickers, wanting to buy their land to clear for grazing (Radachowsky et al., 2012). Lack of trust among the members and greater internal corruption were also present in the concessions that failed. Additionally, these groups were also less educated and had lower-income relative to the other concession households (Fortmann et al., 2017).

While the community forest concessions in the MBR have had their share of ups and downs, on the whole, they have been widely heralded as a story of success, demonstrating how local governance structures can implement market-based mechanisms, such as FSC certification, to achieve their policy goals while promoting environmental sustainability and improve local livelihoods (see https://www.rainforest-alliance.org/articles/guatemala-forest-concessions-global-conservation-model).

Efforts to credibly evaluate empirically the impact of NGOs in improving the causal impact of FSC on deforestation and forest degradation using quantitative data from numerous tropical countries are currently underway.

Ways Forward: The Role of Hybrid Partnerships

A recent couple of decades have experienced a rapid proliferation of voluntary market-based certification schemes that also include land conflict resolution as the requirement for obtaining and retaining certification (e.g., the Roundtable on Sustainable Oil Palm [RSPO] for oil palm production,[4] Bonsucro for sugarcane,[5] and Rainforest Alliance certification for coffee[6]). All of these have requirements pertaining to respecting local rights. We argue that these voluntary market-based interventions have the potential to address land tenure insecurity issues and promote the sustainable use of working forest and other lands in the tropics. Some of these certification schemes are relatively new; it remains to be seen what their impact on local communities is.

Their effectiveness depends on the uptake and continuation of certification, which in turn depends on the demand for certified commodities. While NGOs can help with the administrative, financial, and technical aspects of certification and compliance, communicating the value of the intervention also increases the demand (Polasky et al., 2015). However, rigorous empirical evidence on the impact of certification is still rare, with the most rigorous studies focusing on more convenient locations, where data are easily accessible and are of generally good quality (e.g., Burivalova et al., 2019; Evidensia, 2019: https://www.evidensia.eco/work-with-evidence/visual-summaries/ accessed December 22, 2019).

[4] Principles and Criteria available here: https://rspo.org/resources/certification/principles-criteria/rspo-principle-criteria-certification-systems (accessed Dec. 21, 2019).

[5] Principles and Criteria available here: https://www.bonsucro.com/wp-content/uploads/2017/04/Bonsucro-PS-GDC-English-v4.2.pdf (accessed Dec. 21, 2019).

[6] Principles and Criteria available here: https://www.rainforest-alliance.org/faqs/what-does-rainforest-alliance-certified-mean (accessed Jan. 11, 2020).

While some studies have also considered the social and economic aspects of certification, most focus on environmental outcomes that are easy to quantify (e.g., those based on remote sensing data that allow for large-scale analyses) (Burivalova et al., 2016, 2019; Evisensia: https://www.evidensia.eco, accessed December 22, 2019). Further hampering the communication of the impacts of certification is the inaccessible and highly technical language in which the evaluations of certification are described in academia (Miteva, 2019). Thus, for rigorous evaluation and effective communication of the impacts of certification, multidisciplinary partnerships between businesses, academia, the sustainability standards community, and NGOs are necessary (Miteva, 2019; Milder et al., 2015). These can help ask and answer policy-relevant questions and facilitate the translation of theory onto improvements in the impact of certification on the ground (Miteva, 2019; Milder et al., 2015).

Hybrid partnerships are also needed between formal institutions, specifically, national and international (supra-national) governments, on the one hand, and certifying bodies, on the other. Even though, aided by NGOs and academics, market-based voluntary interventions like FSC forest management certification can in principle be an effective mechanism to mitigate land tenure insecurity when domestic formal and informal institutions fail, they still depend on formal institutions that set the broader governance context nationally (e.g., factors like political stability and corruption, in/out migration policies, and zoning, as well as fiscal incentives like subsidies and taxes for certified products) and internationally (e.g., global markets) (see Box 13.1). That is, for voluntary market-based interventions like FSC to be effective, the incentives set by governments need to align with that of the certification scheme and NGOs (Lambin et al., 2014; Miteva, 2019; see Box 13.1 for an example from Guatemala). While NGOs, through consumer awareness campaigns, can pressure institutions to some extent, the policy coordination cannot be left entirely to the NGO sector.

In conclusion, our review indicates that voluntary market-based interventions like the FSC can and have improved land tenure security in many tropical locations plagued by weak institutions. However, hybrid interventions composed of FSC, NGOs, and formal governments are needed. These should be supported by rigorous science and credible communications in the agenda setting and negotiation, implementation, and monitoring and enforcement phases for the continued impacts in protecting forests on the ground.

References

Adelman, S., & Peterman, A. (2014). Resettlement and gender dimensions of land rights in post-conflict northern Uganda. *World Development, 64*(Dec.), 583–596. https://doi.org/10.1016/j.worlddev.2014.06.031

Alix-Garcia, J., de Janvry, A., & Sadoulet, E. (2005). A tale of two communities: Explaining deforestation in Mexico. *World Development, 33*(2), 219–235. https://doi.org/10.1016/j.worlddev.2004.07.010

Auld, G., Gulbrandsen, L. H., & McDermott, C. L. (2008, October). Certification schemes and the impacts on forests and forestry. *Annual Review of Environment and Resources, 33*(1), 187–211. https://doi.org/10.1146/annurev.environ.33.013007.103754

Baland, J.-M., & Platteau, J.-P. (1996). *Halting the degradation of natural resources*. Food and Agriculture Organization of the United Nations. http://www.fao.org/3/x5316e/x5316e00.htm

Bjørkhaug, I., Bøås, M., et al. (2007). Returning to uncertainty? Addressing vulnerabilities in northern Uganda, Office of the Prime Minister and UNDP. http://www.fafo.no/nyhet/return2uncertanity.pdf

Bocci, C., Fortmann, L., Sohngen, B., & Milian, B. (2018). The impact of community forest concessions on income: An analysis of communities in the Maya Biosphere Reserve. *World Development, 107*(July), 10–21. https://doi.org/10.1016/j.worlddev.2018.02.011

Breukink, G., Levin, J., & Mo, K. (2015). Profitability and sustainability in responsible forestry: Economic impacts of FSC certification on forest operators. WWF. http://d2ouvy59p0dg6k.cloudfront.net/downloads/profitability_and_sustainability_in_responsible_forestry_main_report_final.pdf

Burivalova, Z., Hua, F., Koh, L. P., Garcia, C., & Putz, F. (2016). A critical comparison of conventional, certified, and community management of tropical forests for timber in terms of environmental, economic, and social variables forest certification. *Conservation Letters, 0*(Mar.), 1–11. https://doi.org/10.1111/conl.12244

Burivalova, Z., Allnutt, T. F., Rademacher, D., Schlemm, A., Wilcove, D. S., & Butler, R. A. (2019, June). What Works in Tropical Forest Conservation, and

What Does Not: Effectiveness of Four Strategies in Terms of Environmental, Social, and Economic Outcomes. *Conservation Science and Practice, 1*(6), e28. https://doi.org/10.1111/csp2.28.

Cerutti, P. O., Lescuyer, G., Tacconi, L., Eba'a Atyi, R., Essiane, E., Nasi, R., Tabi Eckebil, P. P., & Tsanga, R. (2017, December 1). Social impacts of the Forest Stewardship Council certification in the Congo basin. *International Forestry Review, 19*(4), 50–63. https://doi.org/10.1505/146554817822295920

Deininger, K., & Castagnini, R. (2006, July). Incidence and impact of land conflict in Uganda. *Journal of Economic Behavior & Organization, 60*(3), 321–345. https://doi.org/10.1016/j.jebo.2004.04.008

Ellis, E. A., Kainer, K., Sierra-Huelsz, J., Negreros-Castillo, P., Rodriguez-Ward, D., & DiGiano, M. (2015). Endurance and adaptation of community forest management in Quintana Roo, Mexico. *Forests, 6*(11), 4295–4327. https://doi.org/10.3390/f6114295

Engel, S., López, R., & Palmer, C. (2006). Community–Industry Contracting over Natural Resource Use in a Context of Weak Property Rights: The Case of Indonesia. *Environmental and Resource Economics, 33*(1), 73–93. https://doi.org/10.1007/s10640-005-1706-5.

Fenske, J. (2011, July). Land tenure and investment incentives: Evidence from West Africa. *Journal of Development Economics, 95*(2), 137–156. https://doi.org/10.1016/j.jdeveco.2010.05.001

Forest Stewardship Council. (2009). Structure, content and local adaptation of Generic Forest Stewardship Standards. FSC-STD20-002 (V3-0) EN, 13 pp.

Forest Stewardship Council. (2015). FSC principles and criteria. https://ic.fsc.org/en/document-center/id/59

Fortmann, L., Sohngen, B., & Southgate, D. (2017, August). Assessing the role of group heterogeneity in community forest concessions in Guatemala's Maya Biosphere Reserve. *Land Economics, 93*(3), 503–526. https://doi.org/10.3368/le.93.3.503

Gómez, I., & Méndez, V. E. (2007). Association of forest communities of Peten, Guatemala. Context, accomplishments and challenges. CIFOR-PRISMA.

Honey-Rosés, J. (2009, October 19). Illegal logging in common property forests. *Society & Natural Resources, 22*(10), 916–930. https://doi.org/10.1080/08941920903131120

Karsenty, A., & Hardin, R. (2017, December 1). Forest concessions in Central Africa: An introduction to the special issue. *International Forestry Review, 19*(4), 1–9. https://doi.org/10.1505/146554817822295975

Lambin, E. F., Gibbs, H. K., Heilmayr, R., Carlson, K. M., Fleck, L. C., Garrett, R. D., De Waroux, P., et al. (2018). The role of supply-chain initiative in reducing deforestation. *Nature Climate Change, 8*(Feb.), 109–116. https://doi.org/10.1038/s41558-017-0061-1

Lambin, E. F., Meyfroidt, P., Rueda, X., Blackman, A., Börner, J., Cerutti, P. O., Dietsch, T., et al. (2014). Effectiveness and synergies of policy instruments for land use governance in tropical regions. *Global Environmental Change, 28*, 129–140. https://doi.org/10.1016/j.gloenvcha.2014.06.007

Larson, A. M., & Soto, F. (2008). Decentralization of natural resource governance regimes. *Annual Review of Environment and Resources, 33*(1), 213–239. https://doi.org/10.1146/annurev.environ.33.020607.095522

Millennium Ecosystem Assessment (MEA) (Program). (2005). *Ecosystems and human well-being*. Washington, D.C: Island Press.

Milder, J. C., Arbuthnot, M., Blackman, A., Brooks, S. E., Giovannucci, D., Gross, L., Kennedy, E. T., et al. (2015, April). An agenda for assessing and improving conservation impacts of sustainability standards in tropical agriculture: Conservation and sustainability standards. *Conservation Biology, 29*(2), 309–320. https://doi.org/10.1111/cobi.12411

Miteva, D. A. (2013). *Three essays on protecting biodiversity in developing countries*. [Order No. 3558969]. Duke University.

Miteva, D. A. (2019, January 7). The integration of natural capital into development policies. *Oxford Review of Economic Policy, 35*(1), 162–181. https://doi.org/10.1093/oxrep/gry029

Miteva, D. A., & Brown, Z. S. (2018). Impact of land tenure insecurity on investment: Evidence from northern Uganda. *Ghanaian Journal of Economics, 6*, 21–42. https://hdl.handle.net/10520/EJC-141f6afa8a

Miteva, D. A., Loucks, C., & Pattanayak, S. K. (2015). Social and environmental impacts of forest management certification in Indonesia. *PLoS ONE, 10*(7), e0129675–e0129675. https://doi.org/10.1371/journal.pone.0129675

Miteva, D. A., Zappernick**, N., Rudulph, J., Brown, Z. S., Ocan, D., Oryema, C., & Kramer, R. A. (2019). Patterns of change in post-conflict forests: Evidence from northern Uganda. *Working Paper*.

Monterroso, I., & Barry, D. (2012). Legitimacy of forest rights: The underpinnings of the forest tenure reform in the protected areas of Petén, Guatemala. *Conservation and Society, 10*(2), 136.

Nittler, J., & Tschinkel, H. (2005). Community forest management in the Maya Biosphere Reserve of Guatemala: Protection through profits. Submitted to the United States Agency for International Development (USAID) and the Sustainable Agriculture and Natural Resource Management (SANREM) Collaborative Research Support Program (CRSP), University of Georgia.

Polasky, S., Tallis, H., & Reyers, B.. (2015). Setting the bar: Standards for ecosystem services, 1–6. https://doi.org/10.1073/pnas.1406490112

Polisar, J., de Thoisy, B., Rumiz, D., Diaz, F. McNab, R. B., Garcia-Anleu, R., Ponce-Santizo, G. Arispe, R., & Venegas, C. (2016, December 1). Using certified timber extraction to benefit jaguar and ecosystem conservation. Ambio. https://doi.org/10.1007/s13280-016-0853-y

Potts, J., Matthew, L., Wilkings, A., Huppé, G., Cunningham, M., & Voora, V. (2014). The state of sustainability initiatives review 2014: Standards and the Green economy. Manitoba, Canada: International Institute for Sustainable Development. https://www.iisd.org/sites/default/files/pdf/2014/ssi_2014.pdf.

Radachowsky, J., Ramos, V. H., McNab, R., Baur, E. H., & Kazakov, N. (2012). Forest concessions in the Maya Biosphere Reserve, Guatemala: A decade later. *Forest Ecology and Management, 268*, 18–28.

Resosudarmo, B. P., Nawir, A. A., Resosudarmo, I. A. P., & Subiman, N. L. (2012). Forest land use dynamics in Indonesia. Australian National University, Arndt-Corden Department of Economics. http://ideas.repec.org/p/pas/papers/2012-01.html

Resosudarmo, I. A. P. (2004, January). Closer to people and trees: Will decentralisation work for the people and the forests of Indonesia? *The European Journal of Development Research, 16*(1), 110–132. https://doi.org/10.1080/0957881041000168876l

Smartwood. (2005). *Forest Management Public Summary for PT Erna Djuliawati*. New York, NY: Rainforest Alliance.

Smartwood. (2006). *SmartWood Certification Assessment Report for PT Intracawood Manufacturing, East Kalimantan, Indonesia*. New York, NY: Rainforest Alliance.

Smartwood. (2007). *Forest Management 2006 CAR Verification Audit Report for PT Sumalindo Lestari Jaya in Samarinda, Kalimantan Timur*. New York, NY: Rainforest Alliance.

Smartwood. (2008). *Corrective Action Request Verification Report for PT Intracawood*. New York, NY: Rainforest Alliance.

Unruh, J., Cligget, L., & Hay, R. (2005, August). Migrant land rights reception and 'clearing to claim' in sub-Saharan Africa: A deforestation example from southern Zambia. *Natural Resources Forum, 29*(3), 190–198. https://doi.org/10.1111/j.1477-8947.2005.00129.x

Vermeulen, C., & Karsenty, A. (2017, December). Towards a Community-Based Concession Model in the DRC. *International Forestry Review, 19*(4), 80–86. https://doi.org/10.1505/146554817822295894.

Watson, E. (2009). Company-led approaches to conflict resolution in the forest sector. The Forest Dialogue. International Institute for Sustainable Development. http://accessfacility.org.s173390.gridserver.com/sites/default/files/G02510.pdf

World Bank. (2009). Post-conflict land policy and administration options: The case of northern Uganda. Washington, DC.

14

Methods to Advance Understanding of Tenure Security: Impact Evaluation for Rigorous Evidence on Tenure Interventions

Kelly W. Jones, Allen Blackman, and Rodrigo Arriagada

Identifying the Causal Impacts of Tenure Security Interventions

Conventional wisdom holds that land tenure security can help achieve a variety of sustainable development goals. Strengthening tenure security is commonly assumed to promote poverty alleviation, agricultural investment and productivity, food security, better health, gender empowerment, and natural resource conservation and restoration (Arnot et al.,

K. W. Jones (✉)
Colorado State University, Fort Collins, CO, USA
e-mail: Kelly.Jones@colostate.edu

A. Blackman
Inter-American Development Bank, Washington, DC, USA

R. Arriagada
Pontificia Universidad Católica de Chile, Santiago, Chile

M. B. Holland et al. (eds.), *Land Tenure Security and Sustainable Development*,
https://doi.org/10.1007/978-3-030-81881-4_14

2011; Holden & Ghebru, 2016; Lawry et al., 2016). Yet rigorous empirical evidence to support these causal links is thin and mixed (Higgins et al., 2018). In fact, even the direction of influence, if any, between tenure security and development objectives has been found to vary. An example is the relationship between tenure security and deforestation (Robinson et al., 2014). On the one hand, stronger secure tenure could spur deforestation by improving land managers' access to credit and commodity markets, which, in turn, could lead to agricultural extensification (Liscow, 2013; Buntaine et al., 2015). But, on the other hand, more secure tenure could stem deforestation by enabling land managers to deter encroachment by outsiders and by dampening their incentives to 'clear land to claim it' (Holland et al., 2017).

Impact evaluation methods aim to assess the causal effects of a project or intervention by comparing what happens when a project, policy, or program is undertaken to what would have happened without that project, policy, or program. Thus, impact evaluation differs from other common forms of monitoring, such as performance measurement, by including a group of observations that do not receive the intervention to better assess what would have happened without it. A well-designed impact evaluation can not only shed light on the average effects of tenure interventions but can also indicate how these impacts are moderated by contextual factors. This can help decision makers understand the conditions under which various types of tenure interventions are likely to be effective. The use of impact evaluation has grown rapidly across development sectors, including agriculture, health, water and sanitation, education, finance, and natural resource management (Cameron et al., 2015). Although the use of impact evaluation methods to study land tenure security interventions has also increased, it lags behind these other sectors. The majority of rigorous evaluations of tenure security have focused on a handful of development outcomes like agricultural investment, access to credit, and income, with fewer measuring outcomes like conflicts, female empowerment, food security, and environmental degradation (Holden & Ghebru, 2016; Higgins et al., 2018).

The gaps in rigorous evidence on the effects of tenure security interventions may be partly due to the fact that tenure evaluations can be more challenging than those for other development interventions, like access to education or health care, for a number of reasons. First, programs and policies aimed at strengthening land tenure often are part of a package of interventions, including, for example, land titling, capacity building, and awareness raising (see Lisher, 2019). In such cases, it can be quite complex and sometimes impossible to disentangle the effects of specific components of this package of interventions. Second, the intended outcomes from strengthening tenure may occur many years after the intervention. Third, as discussed in other chapters, land security is inherently difficult to measure. And finally, outcomes expected from strengthening land tenure are moderated by a multitude of political, economic, and environmental conditions—in general, impacts are heavily dependent on local contextual factors. Given these challenges, evaluating the effect of tenure security interventions requires particular care: clear articulation of a conceptual framework, careful data collection, and attention to the design of a strategy for analyzing those data, so as to identify causal effects.

Articulating a Theory of Change

A theory of change (TOC) is simply a detailed and well-articulated set of hypotheses about exactly how an intervention should affect an intended outcome, that is, it maps out a hypothesized causal chain. Developing a TOC should be the first step in impact evaluation (Qiu et al., 2018). A TOC identifies the logical and ordered causal links moving from the intervention to intended impacts and typically includes a visual illustration (Funnell & Rogers, 2011). These links, visual or otherwise, should be viewed as the hypotheses to be tested in the impact evaluation. A TOC can be based on expert knowledge and/or literature review but should involve local-level or project-level involvement in order to accurately reflect a program's TOC and appropriately account for contextual considerations in tenure interventions. For example, a TOC developed for a land tenure security intervention in Ghana visually details the expected causal links between providing farm-level land tenure documentation to farmers and deforestation-free cocoa production (Persha & Protik, 2019). The TOC consists of a series of 'if-then' statements, laying out the project's central hypothesis that tenure documents will improve land and tree security and improve farmers' access to financial services. The TOC also highlights that the land tenure intervention is only likely to work when contemporaneous development activities occur, in this case, financial and technical services that support shade-grown cocoa and education on land-use planning.

While any TOC can serve as the basis for designing an evaluation, directed acyclic graphs (DAGs) are perhaps particularly well-suited for that purpose because they explicitly recognize multiple types of associations among factors at play (Sills & Jones, 2018; Ferraro & Hanauer, 2015). The generic DAG in Fig. 14.1a shows the causal pathway from the intervention or treatment (T) to final outcome or impact (Y). These impacts could be direct relationships between T and Y, but more often will be propagated through a series of mediators, which correspond to the short-, medium-, or long-term outcomes of the treatment. A mediator (also referred to as a mechanism in the evaluation literature) is a variable that helps explain the relationship between T and Y. In the case of interventions aimed at strengthening tenure security, treatment T might be formalization of land tenure through titling, and the mediator might be changes in tenure security, with the intended impact (Y) of reducing deforestation (Fig. 14.1b). Moderators are distinct from mediators in that they influence the strength of the relationship between T and Y. The researcher typically wants to be aware of these to isolate the casual effect and may be interested in understanding the variation in impact that occurs due to a specific moderator. For example, there may be heterogeneity in the impact of strengthening tenure security by gender, poverty level, or land quality (Fig. 14.1b). By measuring these, the researcher can test for differential effects of an intervention. Confounding factors (sometimes referred to as contextual factors) include any observable or unobservable variables (both of which are ideally included in a DAG or TOC) that mask the true effect of T on Y and need to be controlled for by the researcher in order to measure the 'true' effect of the intervention. In evaluating tenure security, confounding factors may include contemporaneous programs (Fig. 14.1b), or broader socioeconomic or biophysical conditions.

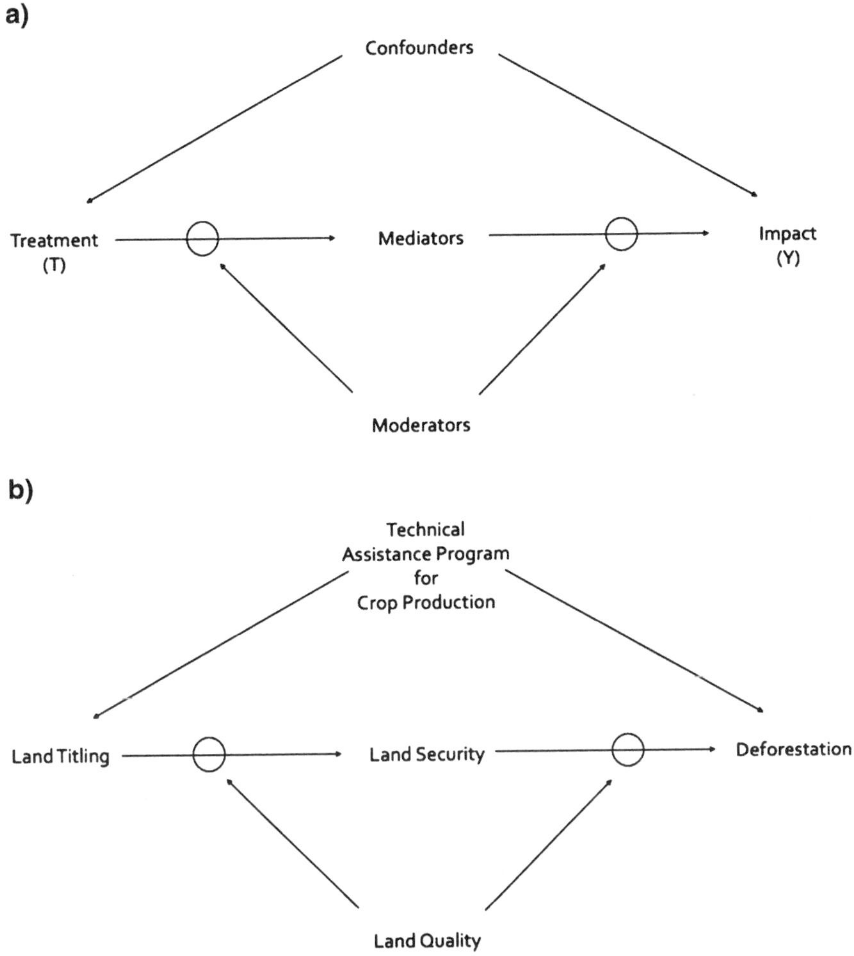

Fig. 14.1 Diagram of **(a)** a generic directed acyclic graph and **(b)** an example of a tenure-related directed acyclic graph

A TOC should also include a 'situation analysis' to explain the current state of insecure conditions and justify why a formal intervention is needed. Additionally, it should lay out the set of assumptions required to move from program inputs to final impacts. In many contexts where tenure interventions are implemented, customary land tenure systems are in place. A situation analysis should highlight current land tenure arrangements (often informal) and how these will be affected (positively or negatively) by formal interventions (Dubois, 1997; Bromely, 2009). These customary systems will interact with and influence how a formal land tenure intervention is implemented and the outcomes of interest (Dubois, 1997; Bromely, 2009).

Collecting Data

A TOC will help to identify the data that need to be collected for an impact evaluation and when those data need to be collected. Regarding the timing of data collection, final impacts may take years to materialize. Therefore, it may be appropriate to start with evaluating intermediate outcomes hypothesized to lead to final impacts. As for what data to collect, at a minimum, the researcher needs to collect data on the intervention, outcomes of interest, and any observable confounding factors that are correlated to both the intervention and the outcomes of interest. So, for example, in the case of an evaluation of land titling on deforestation, data is needed on land titling, deforestation, and confounders such as contemporaneous technical assistance interventions that might affect land titling effectiveness and deforestation decisions (Fig. 14.1b). In addition, in order to test specific mediators or moderators, data also needs to be collected on these factors. To continue with our example, mediators and moderators may relate to the biophysical and socioeconomic characteristics of titled properties.

Using metrics and indicators that are similar to those used in other studies can allow for future comparison and synthesis of evidence of tenure security interventions in systematic reviews. Alternatively, global comparative analyses of similar tenure interventions in different contexts can provide valuable information on how contextual factors influence impacts (e.g., Sunderlin et al., 2018). The use of longitudinal data in impact evaluation is, unfortunately, less common than the use of cross-sectional data, undoubtedly because the costs of collecting longitudinal data are relatively high. However, longitudinal data is much preferred to cross-sectional data because, as discussed below, it facilitates more reliable and more credible analysis. Most evaluations of land tenure interventions are one-off studies (Higgins et al., 2018), providing a static measure of tenure effects. However, most land tenure security interventions lead to a series of intermediate outcomes that are necessary to reach the final impact, and the identification of outcomes in one time period does not necessarily imply that these outcomes will persist over time.

Impact Evaluation: Key Concepts

Conceptually, measuring the effect of an intervention on an intended outcome requires first identifying the group that gets the intervention and then, for that same group, comparing outcomes with the intervention and without it. For example, understanding the effect of land titling on household income requires first identifying the group that receives title and then comparing (i) their income with title and (ii) their income without title. The core challenge in rigorous causal analysis is that of course we cannot observe the outcomes for the group that gets the intervention without the intervention, a scenario termed the counterfactual. In a nutshell, rigorous impact evaluation entails using different strategies for constructing the unobserved counterfactual through use of real-world proxy groups. A number of books and articles discuss these strategies in detail (e.g., Abadie & Cattaneo, 2018; Samii, 2016; Gertler et al., 2016; Imbens & Rubin, 2015; Imbens & Wooldridge, 2009). Here we introduce just a few key concepts and approaches.

Two important types of problems crop up when evaluators construct a counterfactual: selection bias and contemporaneous bias (Ferraro & Pattanayak, 2006; Greenstone & Gayer, 2009). Selection bias becomes an issue when impact evaluations assess outcomes for units of analysis (e.g., households and settlements) that have not received the intervention to represent the counterfactual, but these other units of analysis have systematic differences from the treatment group. For example, a study of the effects of a land titling program on deforestation might use as the counterfactual deforestation on forest management units (FMUs) that were not titled. That is, they might define the effect of titling on deforestation as the difference between the average rates of deforestation on titled versus untitled FMUs. Selection bias would be a concern if the FMUs that were titled were not randomly selected and tended to have pre-existing socioeconomic and/or biophysical characteristics either positively or negatively correlated with deforestation. For example, say, titled FMUs were located closer to cities where deforestation rates were relatively high. In that case, the titled FMUs would very likely have more deforestation than untitled FMUs, but that difference could be due to differences in land-use pressures and completely unrelated to titling. Here, selection bias could result in the misleading conclusion that titling exacerbates deforestation.

Contemporaneous bias occurs when impact evaluations use pre-intervention outcomes for units that received the intervention as the counterfactual. To continue the above example, this counterfactual would be deforestation rates on titled FMUs, but *before* they received titles. In other words, the evaluation relies on a before-after comparison. Contemporaneous bias is a concern when events that coincide with the intervention affect outcomes. So, for our example, this might be a problem if titling coincided with say technical assistance in sustainable forest management or a drop in timber prices over the same period. A before-after comparison could misleadingly conclude that titling reduced deforestation, when it was actually due to technical assistance or a drop in market value of timber. Both selection bias and contemporaneous bias can 'confound' the results of a land tenure security evaluation. Impact evaluation designs typically account for potential confounding due to selection bias and contemporaneous bias by carefully selecting a comparison (or 'control') group to serve as a counterfactual, and by using data on outcomes from both before the intervention ('baseline' data) and after the intervention ('endline' data) (Greenstone & Gayer, 2009; Imbens & Wooldridge, 2009).

Two more terms often encountered in impact evaluation are 'additionality' and 'spillover.' Additionality refers to the changes in outcomes that can be attributed solely to the intervention and not confounding factors. In essence, additionality is the goal of all impact evaluation methods: to be able to determine what effects are because of the project itself over and beyond what would have happened anyway. Spillovers are impacts of the intervention on units not in the treatment group (e.g., Robalino et al., 2017). These impacts could be positive or negative. For example, providing land title to FMUs could cause illegal logging to shift to other nearby FMUs. Spillovers are important to identify because they could contaminate the control group and or influence the estimated program impact.

Impact Evaluation: Key Strategies

Here we provide a brief overview of the principal strategies for rigorous impact evaluation, each of which entails a different approach to constructing the unobserved counterfactual. Experimental designs, or randomized control trials (RCTs), are considered the gold standard of rigorous impact evaluation. They randomly assign an intervention among a population of potential recipients and then define the counterfactual as outcomes for those who did not receive it. The key logic is that if the intervention is randomly assigned, then (if the study population is large enough) the group that does not receive it should have the same average characteristics as the group that did, so that the control group's outcomes are a suitable proxy for the unobserved counterfactual. Put slightly differently, random assignment ensures that selection bias will not be an issue, that is that any correlation between the intervention and the intended outcome will be due to the intervention and not a spurious correlation with confounding factors.

RCTs have been used in a handful of land tenure evaluations (Higgins et al., 2018). While interventions are sometimes assigned to individuals, in the case of tenure security interventions, assignment to clusters or groups (e.g., communities, regions) will often be more politically feasible and more likely to avoid spillovers. Box 14.1 illustrates the use of an RCT assigned at village level to evaluate the impact of a land registration program in Benin.

Box 14.1 An Experimental Study of the Effect of Land Demarcation on Investment

In rural sub-Saharan Africa, only a small portion of farmers have formal titles for the land to which they claim ownership, a situation that in principle could result in under-investment and a low agricultural productivity trap. Existing rigorous evidence on the effect on investment of land tenure interventions aimed at formalizing land tenure is mixed. To help fill that gap, Goldstein et al. (2015) report on results of a randomized controlled trial aimed at identifying the effect of a land formalization program in Benin, the first such randomized controlled trial study. The program they evaluate entailed two key steps, each of which required extensive input from treated communities: first, mapping and demarcating with stone markers all parcels in each treated community and second formally and legally documenting ownership. The formalization program was randomly assigned at the village level and data were collected from individual households within these villages on each of the multiple land parcels claimed by the household. As many as 298 villages participated in the study of which 191 were treated and 98 were not. Secondary baseline and primary endline survey data were compiled or collected for 6064 land parcels claimed by 2972 households in these study villages. The authors develop a formal analytical model that serves as a theory of change. They use OLS regressions to estimate treatment effects. These regressions control for differences in observable characteristics between the treated and control parcels, while randomization is assumed to control for differences in unobserved characteristics. The authors find that two years after the start of implementation, treated households were 23–43 percent more likely to grow perennial cash crops and to invest in trees on their parcels. They also find that household characteristics moderated investment effects. For example, female-headed-treated households invested more in fallowing and tended to shift investment away from demarcated land to less secure parcels in order to secure their claims on them.

Despite their advantage, a number of constraints often make it challenging to implement RCTs in a land tenure context. First, local stakeholders are often justifiably resistant to randomly selecting a group of participants to not receive a tenure security intervention. As a result, RCTs are often used when resources for the intervention in its initial phase are limited so that only a subset of potential units can receive the intervention at one time, so that the subset that gets it initially is randomly selected. Second, in some cases randomization is simply not feasible, e.g., when tenure security interventions are applied to all landholders at regional or national scales. Third, the timing of interventions and outcomes can complicate implementation. RCTs need to be designed before an intervention takes place, which facilitates the collection of baseline data prior to any implementation of the intervention. So RCTs are impractical for interventions already underway.

When random assignment of the treatment and control group are not practical, a quasi-experimental design can be used to estimate the causal effects (e.g., Abadie & Cattaneo, 2018; Gertler et al., 2016). Quasi-experimental designs rely on statistical techniques rather than random assignment of the treatment to construct a valid counterfactual and to estimate the effect of treatment. Here we provide a very brief introduction to the most important causal identification strategies. The first three—matching, difference in difference (DID), and synthetic control method (SCM)—are so-called conditioning strategies. In these designs, one assumes the bias to causal analysis only comes from observable variables or unobservable variables that do not vary over time and equally affects both the treated and control group. Observable variables are those that can be easily measured and controlled for explicitly in statistical analysis (e.g., parcel size), whereas unobservable variables are those where data are nonexistent or too costly to collect (e.g., farmer management capacity). By conditioning on a set of variables, confounding factors can be blocked, and the true relationship between T and Y can be estimated (Fig. 14.1a). The last two methods—regression discontinuity design (RDD) and instrumental variables (IV)—rely on naturally occurring sources of variation in assignment to the intervention. Causal effects are estimated by identifying an exogenous observable variable that is only related to potential outcomes through its effect on the treatment.

The first of the conditioning strategies, matching, entails constructing a counterfactual group by 'matching' treated units to untreated units with similar observable characteristics. Since matching relies on observable data, it is susceptible to bias if important confounding factors are unobservable (i.e., there is no data available) to the researcher. A variety of statistical techniques can be used to help 'match' each treatment unit to observationally similar control units (Abadie & Imbens, 2006; Rosenbaum & Rubin, 1983). One of the most widely used matching algorithms is propensity score matching (PSM). In PSM, a probit or logit regression is used to estimate the probability that each unit in the study sample gets the treatment, conditional on its observed characteristics. The estimated parameters of that regression are then used to predict the probability of receiving the treatment for each unit—the so-called propensity score. Propensity scores can be interpreted as univariate weighted indices of observable characteristics. These scores are then used to match treated and untreated observations. That is, each treated observation is matched to the control observation(s) with the closest propensity score(s). For example, Melesse and Bulte (2015) match households in Ethiopia that received land certification to households that did not based on the following observable variables: age, education, livestock, distance to market, parcel size, distance to parcel, parcel fertility, and parcel slope.

A variety of strategies are available for using matched samples to identify the effects of an intervention. The simplest is to calculate the difference in mean outcomes for the treatment group and matched control group. A more sophisticated method is regression adjustment, that is running a regression using a matched sample that includes treated observations and matched control observations (Imbens & Wooldridge, 2009). Santos et al. (2014) use PSM to match Indian households that were selected for a land-allocation and registration program with households that were not selected. The authors use their matched sample along with a weighted regression analysis to estimate the effect of the program on food security and a number of intermediate outcomes that might influence food security. While matching is one of the easier quasi-experimental methods to employ based on data collection needs, it also requires stronger assumptions and is therefore more susceptible to bias (Ho et al., 2007).

DID, sometimes referred to as before-after-control-impact (BACI), compares before-after changes in outcomes for a treatment group and a control group. Instead of statistically constructing a control group as in matching, this technique uses both the temporal (within) variation and cross-sectional (between) variation to construct the counterfactual outcome. This method can control for selection bias and contemporaneous bias, but only to the extent that these biases apply to both groups in all time periods. DID strategies require data on outcomes from at least two time periods. Economists typically use fixed effects panel regression to estimate the treatment effect in a DID design (Jones & Lewis, 2015).

Box 14.2 illustrates the use of DID in evaluating the impact of a land titling campaign on forest clearing and degradation in indigenous communities in Peru. An increasingly common strategy is to combine matching and DID to reduce potential selection bias. In this case, matching is first used to 'trim' the sample and DID is used to estimate the effect of the intervention on the trimmed sample.

Box 14.2 A Quasi-Experimental Study of the Effect of Titling Indigenous Communities on Forest Cover Change

Blackman et al. (2017) analyze the effect on both forest clearing and degradation of a program in the early 2000s that provided formal legal title to indigenous communities in the Peruvian Amazon. As noted in the main body of this chapter, ex ante the direction of effect of titling on forest cover change is unclear: it could stem clearing by, for example, improving monitoring and enforcement of land-use restrictions, or could spur it, by, for example, improving land managers' access to credit. Blackman et al. develop a detailed theory of change that maps out hypothesized causal links between inputs into titling (community meetings, interactions with external stakeholders, and territorial demarcation), their treatment (the award of title), intermediate outcomes (changes in regulatory pressure, internal governance, public sector interactions, and private sector interactions, livelihoods), and the final outcomes (forest clearing and degradation). Their study sample consists of all 51 communities in the Peruvian Amazon that received title between 2002 and 2005. They use contemporaneous remotely sensed data on their outcomes and GIS data from an indigenous community NGO on the timing and location of the award of title. In addition, they use secondary data from a variety of sources to control for confounding factors (crop prices, rainfall, and temperature) and to test for the effect of moderating factors. They use an 'event-study' quasi-experimental design that entails estimating fixed effects panel data models similar to a DID panel data model except that the regression sample consists only of treated observations, a necessity in this case as GIS data on untitled communities does not exist. In essence, this approach identifies the effect of titling by comparing changes in forest cover before and after title is awarded controlling for observed time-varying confounding factors (crop prices, rainfall, and temperature) as well as all unobserved time-invariant factors. The authors find that titling reduces clearing by more than three-quarters and forest disturbance by roughly two-thirds in a two-year window spanning the year title is awarded and the year afterward. In addition, they find that these effects are more pronounced in communities that are smaller and closer to sizable population centers.

Both matching and DID work best when a large number of units have received the intervention and a large number of potential control units (e.g., households, parcels of land) are available. In that sense, quasi-experimental methods are considered 'data hungry' because we need large samples to find the necessary exogenous identifying variation (Caliendo & Kopeinig, 2008). This requirement poses a challenge for evaluating some types of land tenure interventions, particularly those more likely to have been implemented at large geographic scales, such as a state or country, rather than a household- or parcel-level.

SCM is a conditioning strategy that can be employed for causal analysis when only one or a few units receive the intervention. SCM estimates the effect of an intervention when only a single unit (e.g., one state), or very few units, is exposed to that intervention. It uses both observed characteristics and historical data on outcomes to determine 'similarity' of potential control units. It then estimates a treatment effect by comparing outcomes for the treated unit to a weighted average of outcomes for control units. Thus, it incorporates elements of matching in that it conditions on similar observable characteristics and of DID in that it takes into account changes over time. This method has been promoted as a complement to comparative case studies used in political science, providing a more transparent and robust method of constructing a comparison group (Abadie et al., 2010; Abadie et al., 2015).

RDD relies on the naturally occurring variation in assignment of the intervention to identify a control group. RDD can be used only in situations in which the evaluator can identify a clearly defined threshold for eligibility to receive the intervention that effectively determines assignment into the treatment and control groups (Abadie & Cattaneo, 2018). An example might be a land titling program available only to landholdings at least 10 ha in size. RDD defines treatment and control groups as those just above and below this threshold. The key assumption is that the observable and unobservable characteristics of these groups are the same on average near the threshold of discontinuity (here, households with just under and just over 10 ha of land) so that difference in intended outcomes can be safely attributed to the intervention. This design is considered as good as an experimental approach in terms of controlling for selection bias. A limitation, however, is that the results are only applicable to the sample of the population that is represented at that cut-off point. Ali et al. (2014) apply a spatial RDD in Rwanda to evaluate the impact of a land tenure regularization pilot program, aimed at clarifying rights to land parcels, on gender equality, agricultural investment, and land markets. They use the spatial discontinuity (boundary) between land parcels that were part of the pilot program and land parcels on both sides of the pilot parcels' borders to identify the causal effects.

Finally, an IV design requires that the researcher identify an exogenous variable that affects the treatment but not the potential outcomes, except for its effect through the treatment variable (Angrist et al., 1996). The IV approach controls for selection bias in program assignment by using an instrument (third variable) to represent the treatment variable and thus break the bias that might exist between directly estimating the effect of the treatment on the outcome. A two-step procedure is employed, where the researcher first estimates the effect of the instrument on the treatment and then uses the predicted value of the treatment from this first step to estimate the effect on the outcome of interest. Selection of a 'good' instrument is the most challenging part of an IV approach, and common weaknesses include selecting an instrument that is only weakly correlated with the treatment or is correlated with omitted variables. Similar to RDD, the IV approach can lead to unbiased estimates of the treatment effect, but only for the sample that is represented by the exogenous variable. In a systematic review by Higgins et al. (2018), the IV approach was the most common quasi-experimental design used to measure the impact of tenure security on development outcomes. Common instruments used for tenure security interventions included a household's distance to land offices or parcels, mode of land acquisition, exposure to information on reforms, and tenure status of neighboring parcels (Higgins et al., 2018).

All of the evaluation methods discussed above have their strengths and weaknesses, and each of the quasi-experimental approaches comes with a set of assumptions that have to be met in order to effectively measure the 'true' impact. A well-articulated TOC can help identify when the assumptions of each method are likely to be met. Combining any of the above methods with qualitative data from focus groups, interviews, or ethnographic methods can also help elucidate the various causal pathways, or mediators, that are operating; the effect of moderators on heterogeneity in outcomes; and the influence confounding factors might be having on masking impacts. Box 14.3 illustrates how qualitative information can be used alongside impact evaluation to shed light on effects, or lack thereof, of tenure interventions and reveal unintended outcomes. In the context of participatory research, in which the researcher returns to the field to clarify questions and resolve anomalies, qualitative observations can be combined with quantitative data to elicit knowledge about motivations to participate in specific programs and to verify estimated impacts (see Rao & Woolcock, 2003 and Arriagada et al. 2009).

Box 14.3 A Mixed Methods Evaluation of the Impact of Land Titling on Deforestation

Holland et al. (2017) combine the quasi-experimental impact evaluation method of DID with focus group discussions to elucidate the causal pathways (i.e., mediators) of a land titling intervention in Ecuador. Holland et al. evaluate the impact of receiving formal land titles in 2009 on avoided deforestation around a national forest reserve in the Ecuadorian Amazon. They compare 1067 land parcels that receive a formal land title to 268 land parcels that do not receive a formal land title, but that are located in the same 61 communities. The authors use data on annual changes in forest cover between the years 2000 and 2014 and estimate the effect of titling on avoided deforestation using fixed effects panel regression methods. They find that the average effect of receiving a formal land title is that deforestation decreases by 37 percentage points. They combined this analysis with 15 community focus group discussions. Through focus group discussion they learn that *de facto* tenure security is already strong before receiving formal land titles, a finding that has been documented in other studies of rural communities in Ecuador. Individuals were asked about specific mechanisms or pathways through which land titling had changed their forest behaviors, but respondents gave no indication that titles had led to abatement of forest harvesting. Instead, the impact of land titling appeared to occur only because pressure on lands left untitled spiked during the evaluation time period. It is not clear if this spike was due to illegal activity or other pressures. Unintended consequences of the land titling campaign were also discovered in the focus group interviews since many respondents were dissatisfied by restrictions put on subdividing their land for future generations. These nuances to the impact of formal land titles on forest conservation would not have been discovered without the use of mixed methods.

For all of the approaches described above, it is important to consider the internal and external validity of the results. Results are said to be internally valid when the evaluation has controlled for biases, including selection bias, contemporaneous bias, and bias due to spillovers. Internal validity is what impact evaluation designs strive for—estimating the true impact of the program for that population. Results are said to be externally valid when they can be generalized to the population of all eligible units. External validity can be harder to achieve in impact evaluation. It requires that the treatment and control groups be representative of the population of all eligible units. Certain research designs lend themselves better to this, in particular those that randomly assign treatment and control units from the population of eligible units. Others are more problematic in this regard: IV and RDD. Measuring potential moderators and estimating sub-group impacts, in addition to average impacts, can increase the ability to extrapolate results to all eligible units (Sills & Jones, 2018).

Advancing Our Understanding of Tenure Security

The growing attention to the potential benefits of strengthening tenure security on sustainable development outcomes is a welcome advance. Evidence on where, when, and why these impacts are occurring would help ensure more efficient use of financial investments. As noted, tenure security interventions and their links to intended final impacts are often complicated, making measurement and isolating the impact of tenure security alone difficult. This underscores the need for well-designed impact evaluations. That, in turn, requires developing a TOC for interventions that clearly articulates the links between interventions and final outcomes and that identifies confounding factors, moderators, and potential mediators. It also requires collecting requisite data and selecting experimental and quasi-experimental impact evaluation methods that, to the extent possible, will control for selection bias, contemporaneous bias, and spillover.

While the number of impact evaluations of tenure security interventions is increasing, the focus is limited to a small set of potential outcomes (Higgins et al., 2018). More research efforts are needed that measure the effects of tenure security on sustainable development goals such as gender equality, public health, and environmental degradation. Additionally, impact evaluations of tenure security interventions would benefit from prioritizing comparative and longitudinal studies. Comparative studies can ensure similar metrics are used in multiple contexts, shedding light on where tenure interventions work and why. Longitudinal studies would help advance knowledge on the dynamics of strengthening land tenure and take into account how exogenous changes may influence the sustained impacts of secure tenure. Any increase in the evidence base on land tenure needs to acknowledge multiple ways of knowing (Tengö et al., 2014) and should ensure that research questions and assessments are co-developed with the practitioners and decision makers that have on the ground understanding of the local tenure context.

It is important to note that not all tenure security interventions will be conducive to rigorous impact evaluation. Evaluations may not be feasible where project conditions are not generalizable to larger areas and where the costs of conducting an evaluation are high. But not all tenure interventions need to be evaluated to advance our knowledge on what works, where, and why. Increasing the use of impact evaluation methods to assess tenure security interventions by even a fraction would greatly enhance the evidence base and provide decision makers with better knowledge on what interventions will work in different contexts.

References

Abadie, A., & Cattaneo, M. D. (2018). Econometric methods for program evaluation. *Annual Review of Economics, 10*, 465–503.

Abadie, A., Diamond, A., & Hainmueller, J. (2015). Comparative politics and the synthetic control method. *American Journal of Political Science, 59*(2), 495–510.

Abadie, A., Diamond, A., & Hainmueller, J. (2010). Synthetic control methods for comparative case studies: Estimating the effect of California's tobacco control program. *Journal of the American Statistical Association, 105*(490), 493–505.

Abadie, A., & Imbens, G. W. (2006). Large sample properties of matching estimators for average treatment effects. *Econometrica, 74*(1), 235–267.

Ali, D. A., Deininger, K., & Goldstein, M. (2014). Environmental and gender impacts of land tenure regularization in Africa: pilot evidence from Rwanda. *Journal of Development Economics, 110*, 262–275.

Angrist, J. D., Imbens, G. W., & Rubin, D. B. (1996). Identification of causal effects using instrumental variables. *Journal of the American Statistical Association, 91*(434), 444–455.

Arnot, C. D., Luckert, M. K., & Boxall, P. C. (2011). What is tenure security? Conceptual implications for empirical analysis. *Land Economics, 87*(2), 297–311.

Arriagada, R., Sills, E., Pattanayak, S. K., & Ferraro, P. (2009). Combining qualitative and quantitative methods to evaluate participation in Costa Rica's Program of Payments for Environmental Services. *Journal of Sustainable Forestry, 28*(3-5), 343–367.

Blackman, A., Corral, L., Lima, E., & Asner, G. (2017). Titling Indigenous Communities Protects Forests in the Peruvian Amazon. *Proceedings of the National Academy of Sciences of the United States of America, 114*, 123–4128.

Bromely, D. W. (2009). Formalising property relations in the developing world: the wrong prescription for the wrong malady. *Land Use Policy, 26*(1), 20–27.

Buntaine, M. T., Hamilton, S. E., & Millones, M. (2015). Titling community land to prevent deforestation: an evaluation of a best-case program in Morona-Santiago, Ecuador. *Global Environmental Change, 33*, 32–43.

Caliendo, M., & Kopeinig, S. (2008). Some Practical Guidance for the Implementation of Propensity Score Matching. *Journal of Economic Surveys, 22*(1), 31–72.

Cameron, D. B., Mishra, A., & Brown, A. N. (2015). The growth of impact evaluation for international development: how much have we learned? *Journal of Development Effectiveness, 8*(1), 1–21.

Dubois, O. (1997). Rights and Wrongs of Rights to Land and Forest Resources in Subsaharan Africa: Bridging the Gap between Customary and Formal Rules. IIED Forest Participation Series No. 10. IIED, London, UK.

Ferraro, P. J., & Hanauer, M. M. (2015). Through what mechanisms do protected areas affect environmental and social outcomes? *Phil. Trans. R. Soc. B, 370*(20140267), 1–11.

Ferraro, P. J., & Pattanayak, S. K. (2006). Money for nothing? A call for empirical evaluation of biodiversity conservation investments. *PLoS Biology, 4*(4), 1.

Funnell, S. C., & Rogers, P. J. (2011). *Purposeful program theory: Effective use of theories of change and logic models*. Jossey-Bass.

Gertler, P. J., Martinez, S., Premand, P., Rawlings, L. B., & Vermeersch, M. J. (2016). *Impact Evaluation in Practice* (2nd ed.). World Bank.

Goldstein, M., Houngbedji, K., Kondylis, F., O'Sullivan, M., & Selod, H. (2015). Formalization without certification? Experimental evidence on property rights and investment. *Journal of Development Economics, 132*, 57–74.

Greenstone, M., & Gayer, T. (2009). Quasi-experimental and experimental approaches to environmental economics. *Journal of Environmental Economics and Management, 57*(1), 21–44.

Higgins, D., Balint, T., Liversage, H., & Winters, P. (2018). Investigating the impacts of increased rural land tenure security: a systematic review of the evidence. *Journal of Rural Studies, 61*, 34–62.

Ho, D. E., Imai, K., King, G., & Stuart, E. A. (2007). Matching as nonparametric preprocessing for reducing model dependence in parametric causal inference. *Political Analysis, 15*(3), 199–236.

Holden, S. T., & Ghebru, H. (2016). Land tenure reforms, tenure security and food security in poor agrarian economies: casual linkages and research gaps. *Global Food Security, 10*, 21–28.

Holland, M. B., Jones, K. W., Naughton-Treves, L., Freire, J. L., Morales, M., & Suárez, L. (2017). Titling land to conserve forests: the case of Cuyabeno Reserve in Ecuador. *Global Environmental Change, 44*, 27–38.

Imbens, G. W., & Rubin, D. B. (2015). *Causal Inference in Statistics and the Medical and Social Sciences*. Cambridge University Press.

Imbens, G. W., & Wooldridge, J. M. (2009). Recent Developments in the Econometrics of Program Evaluation. *Journal of Economic Literature, 47*(1), 5–86.

Jones, K. W., & Lewis, D. J. (2015). Estimating the counterfactual impact of conservation programs on land cover outcomes: The role of matching and panel regression techniques. *PLoS ONE, 10*(10), e0141380.

Lawry, S., Samii, C., Hall, R., Leopold, A., Hornby, D., & Mtero, F. (2016). The impact of land property rights interventions on investment and agricultural productivity in developing countries: a systematic review. *Journal of Development Effectiveness, 9*(1), 61–81.

Liscow, Z. D. (2013). Do property rights promote investment but cause deforestation? Quasi-experimental evidence from Nicaragua. *Journal of Environmental Economics and Management, 65*, 241–261.

Lisher, J. W. (2019). *Guidelines for impact evaluation of land tenure and governance interventions*. UN Human Settlements Programme.

Melesse, M. B., & Bulte, E. (2015). Does land registration and certification boost farm productivity? Evidence from Ethiopia. *Agricultural Economics, 46*, 757–768.

Persha, L., & Protik, A. (2019). Evaluation of the "Supporting Deforestation Free Cocoa in Ghana" Project Bridge Phase: Evaluation Design Report. USAID, Washington, DC. Accessed at https://www.land-links.org/wp-content/uploads/2019/05/Evaluation-of-the-Supporting-Deforestation-Free-Cocoa-in-Ghana-Project-Bridge-Phase-Evaluation-Design-Report.pdf

Qiu, J., Game, E. T., Tallis, H., Olander, L. P., Glew, L., Kagan, J. S., Kalies, E. L., Michanowicz, D., Phelan, J., Polasky, S., Reed, J., Sills, E. O., Urban, D., & Weaver, S. K. (2018). Evidence-based causal chains for linking health, development and conservation actions. *BioScience, 68*(3), 182–193.

Rao, V., & Woolcock, M. (2003). Integrating qualitative and quantitative approaches in program evaluation. In F. Bourguignon & L. P. da Silva (Eds.), *The impact of economic policies on poverty and income distribution: Evaluation techniques and tools* (pp. 165–190). Oxford University Press.

Robalino, J., Pfaff, A., & Villalobos, L. (2017). Heterogeneous local spillovers from protected areas in Costa Rica. *Journal of the Association of Environmental and Resource Economists, 4*(3), 795–820.

Robinson, B. E., Holland, M. B., & Naughton-Treves, L. (2014). Does secure land tenure save forests? A meta-analysis of the relationship between land tenure and tropical deforestation. *Global Environmental Change, 29*, 281–293.

Rosenbaum, P. R., & Rubin, D. B. (1983). The central role of the propensity score in observational studies for causal effects. *Biometrika, 70*(1), 41–55.

Santos, F., Fletschner, D., Savath, V., & Peterman, A. (2014). Can government-allocated land contribute to food security? Intrahousehold analysis of West Bengal's Microplot Allocation program. *World Development, 64*, 860–872.

Samii, C. (2016). Causal empiricism in quantitative research. *The Journal of Politics, 78*(3), 941–955.

Sills, E. O., & Jones, K. W. (2018). Causal inference in environmental conservation: the role of institutions. *Handbook of Environmental Economics, 4*, 395–437.

Sunderlin, W. D., de Sassi, C., Sills, E. O., Duchelle, A. E., Larson, A. M., Resosudarmo, I. A. P., Awono, A., Kewka, D. L., & Huynh, T. B. (2018). Creating an appropriate tenure foundation for REDD+: The record to date and prospects for the future. *World Development, 106*, 376–392.

Tengö, M., Brondizio, E. S., Elmqvist, T., Malmer, P., & Spierenburg, M. (2014). Connecting diverse knowledge systems for enhanced ecosystem governance: the multiple evidence base approach. *Ambio, 43*(5), 579–591.

15

Emerging Research Needs and Policy Priorities for Advancing Land Tenure Security and Sustainable Development

Yuta J. Masuda, Brian E. Robinson, Margaret B. Holland, Tzu-Wei Joy Tseng, and Alain Frechette

Significant Advances But Gaps Remain

The evidence on the importance of land tenure security (LTS) in sustainable development is increasingly clear: research continually highlights the critical role of tenure security in biodiversity conservation (Díaz et al., 2019; Erbaugh et al., 2020; Garnett et al., 2018), climate change mitigation and resilience (Intergovernmental Panel on Climate Change (IPCC), 2020), poverty reduction (Besley & Burgess, 2000; Deininger, 2003),

Y. J. Masuda (✉)
Global Science, The Nature Conservancy, Arlington, VA, USA
e-mail: ymasuda@tnc.org

B. E. Robinson
Department of Geography, McGill University, Montreal, QC, Canada

M. B. Holland
Department of Geography & Environmental Systems, University of Maryland, Baltimore County, Baltimore, MD, USA

M. Holland et al. (eds.), *Land Tenure Security and Sustainable Development*,
https://doi.org/10.1007/978-3-030-81881-4_15

women's empowerment (Meinzen-Dick et al., 2019; O'Sullivan, 2017), and many other topics. Recent systematic reviews (Fenske, 2011; Higgins et al., 2018; Lawry et al., 2017; Meinzen-Dick et al., 2019; O'Sullivan, 2017; Robinson et al., 2014; Tseng et al., 2021)—as well as the many chapters in this book—have highlighted the significant steps forward in our understanding of how LTS impacts environmental and human well-being outcomes, but these have also underscored significant research gaps that still remain. Perhaps the clearest message to emerge from decades of research on LTS is that the determining factors (contextual and otherwise) are complex and multilayered (Robinson et al., 2018; van Gelder, 2010), and that titling alone is not a panacea, nor likely a sufficient stand-alone strategy, for addressing tenure insecurity (Sjaastad & Cousins, 2009) (see also Chap. 11 in this volume). The complexity of tenure insecurity has been illustrated in many in-depth studies focusing on specific contexts or subpopulations (Holland et al., 2014, 2017; Naughton-Trevesa et al., 2011; Orellano et al., 2015). In short, historical injustices, failed attempts to remedy them, and policies designed to entice new migrants to areas, as well as other factors, all have contributed to a web of related but distinct factors driving tenure insecurity. Chapter 2 provides a succinct summary of the complicated and contentious history of land rights, and how LTS for much of the world today has been built on inequitable access and control of land, where elites have often held control over who gets rights to the land.

The landscape of stakeholders engaging on tenure security issues has also increased in recent decades. Environmental, civil society, and other non-governmental actors are playing an increasingly important role in raising awareness about, and, when possible, strengthening the tenure security of women, smallholder farmers, indigenous groups, and other subpopulations. Government actors at all levels also play a pivotal role in

T.-W. J. Tseng
Institute for Health and Social Policy, McGill University, Montreal, QC, Canada

A. Frechette
Rights and Resources Initiative, Washington, DC, USA

ensuring equitable and transparent LTS. Chapters 2 and 11 highlight the uneven role governments have had in securing tenure for their people, but nonetheless they occupy a critical position in determining how LTS is addressed. A primary challenge is that securing tenure tends to be a zero-sum game—that is, securing tenure for one group may come at the expense of another group (Adam, 2020; Natcher et al., 2009). Activists and civil groups (e.g., Chap. 12) also play a role in securing tenure across the world, as they raise awareness within and outside countries on land rights issues. A common challenge for all stakeholders, however, is that without rigorous evidence of the causes and consequences of LTS, uncertainty will cloud policy priorities and strategies. Research and policy must still make significant advances if efforts to secure tenure across the world are to be successful.

An Urgent Need to Expand the Breadth and Depth of Studies on LTS Impacts

A recent systematic review by Tseng et al. (2021) of more than one hundred studies found strong support for a positive relationship between LTS on human well-being outcomes, such as agricultural investments, increased credit, women's empowerment, and food security. For environmental outcomes, there is generally positive support for the effects of LTS on forest conditions or conservation investments. But context and nuance matter in the strength and direction of LTS' effect on human well-being and environmental outcomes. Critically, the strength and direction of LTS is influenced by the bundle of rights associated with a given tenure system and the myriad social, economic, political, and environmental factors (both internal and external to communities) that condition the *de facto* performance of such arrangements. For instance, Robinson et al. (2014) reviewed 118 cases and found tenure security is a necessary but insufficient condition for incentive-based forest policies. Additionally, if we accept LTS is a basic covenant of most sustainable social-ecological relationships, then the adequacy of associated arrangements (social, political, and economic institutions) in supporting positive social and environmental outcomes becomes critical. Indeed, it is also these very factors that can directly influence perceived tenure security (van Gelder, 2010). As such, syntheses of existing evidence shed light on what we still *do not know* about the role LTS plays in environmental and human well-being outcomes, pointing to critical research directions needed to advance the design and implementation of evidence-informed policies around strengthening LTS.

First, few studies have rigorously examined the effect of LTS on both human well-being *and* environmental outcomes simultaneously, making it difficult to understand potential tradeoffs or synergies. Tseng et al. (2021) found only 20% of 117 studies in their sample attempted to estimate the causal effect of LTS simultaneously for human well-being and environmental outcomes. Second, studies often lacked the time scales necessary to estimate long-run effects on environmental outcomes, raising questions about whether impacts from strengthening LTS have enduring positive effects on biodiversity and ecosystem services. Instead, studies frequently examined actions taken by landholders that could plausibly lead to improved environmental outcomes, such as investments in soil (Deininger et al., 2011) or forest conservation (Holland et al., 2017). While suggestive, it is often unclear how long those practices endure and thus lead to the desired outcomes. Third, there is a need to study the effects of LTS in a more diverse set of biomes. The majority of studies largely examined effects of LTS on tropical forests—perhaps one of the easiest to measure environmental change over longer time periods given the wide-scale availability of remotely sensed forest cover data (Hansen et al., 2013)—or on modified lands, such as farms. Grasslands, wetlands, deserts, and even dry forests need further study, as land use pressures can differ compared to forests and agricultural lands. Fourth, most rigorous studies are also concentrated in a few countries, and there is overrepresentation in Ethiopia and China which have unique land laws and administration systems, making comparison across these contexts challenging. Fifth, a large proportion of studies have evaluated the impacts of LTS on economic outcomes, while other areas of human well-being remain understudied.

Finally, more study is needed to evaluate how climate change, demographic transitions, migration, and other macro-factors affect LTS over longer time horizons. Consider, for instance, that rural to urban migration continues to far outpace urban to rural migration (United Nations Department of Economic and Social Affairs, 2019), raising questions about who will own, live on, and manage rural lands in the future. Changes in family size may also have significant implications for how land is allocated: population growth in some parts of the world may intersect with inheritance laws, leading to smaller and smaller plots of land. For example, expected population growth in Sub-Saharan Africa may require dividing up already-small farms even further. Climate change further complicates these dynamics, rendering some lands uninhabitable and triggering temporary or even permanent displacement of people, leading to significant migration to areas more resilient to the effects of climate change. This may amplify land use pressures and risk food security in source or destination lands (as explored in Chap. 6).

From a policy perspective, rigorous studies have largely focused on evaluating the impacts of legal titling (as reviewed in Chap. 11), thus raising questions about the efficacy of other policy tools, or the effects of a combination of different efforts (e.g., combining titling, awareness raising, and capacity building). Tseng et al. (2021) identified significant research gaps for interventions that raise awareness about land rights and capacity building interventions (e.g., increasing administrative capacity of local governments). A serious practical, logistical, as well as research, challenge is that multiple stakeholders often champion different factors affecting LTS. For instance, an environmental non-governmental organization may work with a community and outside stakeholders to create land use plans, thus informally increasing recognition of community lands with those outside the community. At the same time, however, local authorities may provide private land titles to community lands, thus creating countervailing efforts that undermine the LTS of the community. Understanding factors that drive effective collaboration amongst stakeholders is crucial to achieve LTS for the target population.

There is also a need to study policy implementation because LTS policies can be complex. For instance, formalization of land ownership through legal titles may require raising awareness about titling efforts, systems to handle paperwork and administrative processes, identifying plots of land and resolving any conflicts over ownership, and resolving any mismatches between customary rights that may conflict with the statutory system. The logistics, institutional capacity, and costs involved are not trivial (Notess et al., 2020). Along this pathway, numerous factors can create inefficiencies or policy implementation failures. If literacy is low, efforts to raise awareness may be hindered. There may be inequitable land ownership if local patriarchal systems are dominant, even in cases where national laws indicate women are allowed to own land. Community leaders and other officials must have the know-how, capacity, and legitimacy to resolve any conflicts over land ownership. Any mismatches in statutory and customary systems must also be resolved (e.g., how should community land be legally recognized if titles are only given to individuals?). Careful documentation and study of the various implementation challenges of policy efforts are needed to create robust policies that focus on equity in strengthening LTS.

More work is also needed to evaluate how policies aiming to strengthen LTS may lead to uneven distributional impacts across contexts and for different groups of people. Women, Indigenous Peoples and traditional local communities (IPLCs), recent migrants, and other groups may often have less political and economic power to engage in processes that can, for instance, secure their land rights or resolve conflicts. Earlier chapters in this volume explore how LTS and policies affecting it can impact IPLCs (Chaps. 4 and 12) and women (Chap. 5) can be weakened or strengthened. Without an understanding of inequities that result from policy design and implementation, even well-intentioned efforts will fail to address these issues.

Methodologically, several issues must be advanced, although we focus on one aspect that has emerged in our review as especially important. While there have been significant conceptual advances for understanding LTS (Robinson et al., 2018; Simbizi et al., 2014; van Gelder, 2010)—namely that perceptions of LTS in particular are important—how LTS is measured in studies analyzing its effects remains fractured. Tseng et al. (2021) found little overlap in how LTS is measured, with an overwhelming number of studies measuring land rights, and only a few studies measured landholders' subjective perceptions. Land rights themselves can be complex to measure since, for example, *de jure* and *de facto* tenure regimes may include different or sometimes overlapping bundles of rights

(Chap. 3). Notably, studies that examine the effects of the same policy were found to use different measures of LTS, sometimes even when using the same dataset. How LTS is measured and evaluated in studies has implications for our understanding of the causes and consequences of LTS. An intervention that has been successfully implemented may have little effect on the population's perceived security, especially if overall incentives (whether from internal or external sources) do not change for the parties involved. Comparing results between studies may be difficult or impossible, and tracking progress on increasing LTS may be challenging if different actors are measuring LTS in their own way.

Testing New Policies to Accelerate Secure Tenure Across the World

Despite the increasing number of actors addressing tenure insecurity across the world, there is still a need to develop and test new interventions if LTS is to have its expected impact for advancing global sustainability goals, such as Sustainable Development Goal 1.4.2 and 5.a.1. Tseng et al. (2021) found policies aiming to strengthen LTS often involve several interventions implemented by multiple actors. But most interventions in this review stemmed from macro-economic policy directives and tended to be implemented through a top-down approach, with little or no input from the target population and community. Other efforts by the Tenure Facility, the Rights and Resources Initiative, and others are actively supporting bottom-up actions and may result in quite different outcomes and dynamics. A first step may be to invest in careful evaluations of these programs, distinguishing between their origins, and disentangling the marginal benefit of the various interventions in strengthening LTS. It is likely that factors, such as the legitimacy and trust of governments and community leaders that may be implementing the intervention, will influence the efficacy of the policy, and different combinations of policies could yield more promising outcomes.

Technological innovations provide a promising pathway for increasing the efficiency and transparency of land rights. For instance, the Cadasta

Foundation has been testing digital tools to create, manage, and store data on land rights and property boundaries. Blockchain-based land titles have been pilot tested in Georgia (Shang & Price, 2018) and elsewhere to overcome challenges of maintaining and updating reliable land registries. However, as technologies are tested and rolled out, it is critical to increase the capacity of landholders, as well as those seeking to own land, to understand laws and technological tools to ensure equitable access and use of such resources. To be clear, these technological improvements may help reduce logistical and administrative burdens, but the hard work of reconciling land disputes or other underlying social frictions that so often underlie the lack of clarity around land boundaries still remain. These are difficult barriers that must be addressed through conflict resolution, mediation, or restitution and compensation when appropriate.

Moving Forward, Quickly

From their inception, the sustainable development goals (SDGs) established an ambitious set of global goals to be achieved by 2030. Nearly half of the SDGs rely—directly or indirectly—on strengthening LTS, highlighting the urgent need to accelerate efforts to strengthen LTS across the world if we are to advance development outcomes, preserve nature, and mitigate and adapt to the effects of climate change. There are several areas for immediate collaboration and action.

First, wishful thinking and assumptions about what works and what does not work should be avoided and replaced with evidence-informed decision-making. More research is generally needed to uncover causal mechanisms and the factors that mediate the efficacy of particular policies and programs, but the research must be rigorous, done at larger scales, and examine LTS effects on both human well-being and environmental outcomes over longer time horizons. Second, more work is needed to examine the implementation of various policies. Third, more work should focus on gathering landholder perceptions about interventions—ranging from their design, their implementation, and the effects on the outcomes they care about. Finally, greater collaboration and information sharing is necessary. A diverse and vast set of stakeholders are now working to strengthen LTS, and they must be willing and open to share information about the challenges that emerge, when policies fail, when policies succeed, and report on unintended consequences. A critical step may be to unify measurement across independent researchers so that studies can be compared across contexts and over time, including urban and rural contexts. Sustained and collaborative dialogue between practitioners, rights-holders, and researchers working on LTS issues is imperative. Cross-fertilization, sharing of information, and joint data collection and analysis among practitioners and rights-holders is critical, especially with those based in the Global South. This knowledge can inform policy and practice *in situ*, especially by rights-holder organizations that are leading efforts for the recognition of their tenure rights.

Recent work indicates practitioners and researchers working on LTS seemingly characterize LTS differently (Masuda et al., 2020), suggesting information exchange between researchers and practitioners may be limited. Efforts should also continue to encourage peer-to-peer learning within and among practitioner groups. More attention on the questions and issues that communities and rights-holders face could also better inform the next generation of policies. This could include feedback on perceived outcomes, what works (and does not) in implementing land tenure interventions, understanding how tenure reforms can be accelerated, and how external actors and resources can best support those processes.

It is also necessary to bridge research, practice, and policy across disciplines, fields, and sectors. The land system science community has focused on landscape-scale changes in land and human well-being outcomes, often with a lens on governance. Discourse in that field is beginning to explore the importance of land rights within a general land governance perspective, but sometimes without explicitly talking about *land tenure security*. Different use of terms results in a lack of dialogue between these realms of research and practice (McSweeney & Coomes, 2020).

Another key to improving research, practice, and policy is in understanding the connections between LTS issues for rural communities and urban populations. This seems especially important with rising cases of conflicts, economic migration, agro-industrial land acquisitions, and climate change. Our understanding of the concept of LTS and the factors that support it could change substantively over the coming decades. As we understand more about the dynamics of rural-urban migration, in particular its bi-directional and cyclical characteristics, we see these as areas of work that should be further integrated.

Policymakers and funders should invest more, over longer time periods, to understand and disseminate what policy mechanisms are most likely to be successful. They should also be open to experimenting with new tools—whether they are technical or behavioral nudge interventions—and strongly signal to implementing actors that failure is expected but should also be openly shared and remedied. Generally, these actors have enormous influence in shaping the conversation around the set of actions that are taken, and can dictate where and what will be done through their funding decisions. Donors, multilateral institutions, and development institutions have invested comparatively little money until now in LTS while acknowledging that most of their desired development outcomes hinge on the realization of LTS at scale. There is growing awareness of this disconnect, as demonstrated by initiatives like the World Bank's Forest Carbon Partnership Fund and increasing public discussions around these issues. To this end, initiatives to build multilateral funding and support are critical. One example being led by the Rights and Resources Initiative is a new global initiative called Path to Scale that aims to raise global ambition and funding in support of LTS. Support is being provided by donors such as the Foreign, Commonwealth & Development Office of the UK (FCDO—formerly DFID), the German Federal Ministry for Economic Cooperation and Development (BMZ), and the Swedish Institute for Development Assistance (SIDA), and private foundations. Still, much more needs to happen, especially in terms of getting technical and financial resources to community actors and leaders directly.

References

Adam, A. G. (2020). Understanding competing and conflicting interests for peri-urban land in Ethiopia's era of urbanization. *Environment and Urbanization, 32*(1), 55–68. https://doi.org/10.1177/0956247819890215

Besley, T., & Burgess, R. (2000). Land reform, poverty reduction, and growth: Evidence from India. *Quarterly Journal of Economics, 115*, 389–430. https://doi.org/10.1162/003355300554809

Deininger, K. (2003). *Land policies for growth and poverty reduction* (Vol. 7). World Bank Policy Research Report. https://doi.org/10.1007/s00784-003-0213-8

Deininger, K., Ali, D. A., & Alemu, T. (2011). Impacts of land certification on tenure security, investment, and land market participation: Evidence from Ethiopia. *Land Economics, 87*(2), 312–334. https://doi.org/10.3368/le.87.2.312

Díaz, S., Settele, J., Brondízio, E., Ngo, H. T., Guèze, M., Agard Trinidad, J., et al. (2019). *IPBES, summary for policymakers of the global assessment report on biodiversity and ecosystem services of the Intergovernmental Science-Policy Platform on Biodiversity and Ecosystem Services.* Intergovernmental Science-Policy Platform on Biodiversity and Ecosystem Services.

Erbaugh, J. T., Pradhan, N., Adams, J., Oldekop, J. A., Agrawal, A., Brockington, D., et al. (2020). Global forest restoration and the importance of prioritizing local communities. *Nature Ecology and Evolution, 4*(11), 1472–1476. https://doi.org/10.1038/s41559-020-01282-2

Fenske, J. (2011). Land tenure and investment incentives: Evidence from West Africa. *Journal of Development Economics, 95*(2), 137–156. https://doi.org/10.1016/j.jdeveco.2010.05.001

Garnett, S. T., Burgess, N. D., Fa, J. E., Fernández-Llamazares, Á., Molnár, Z., Robinson, C. J., et al. (2018). A spatial overview of the global importance of Indigenous lands for conservation. *Nature Sustainability, 1*(7), 369–374. https://doi.org/10.1038/s41893-018-0100-6

Hansen, M. C., Potapov, P. V., Moore, R., Hancher, M., Turubanova, S. A., Tyukavina, A., et al. (2013). High-resolution global maps of 21st-century forest cover change. *Science, 345*, 850–853.

Higgins, D., Balint, T., Liversage, H., & Winters, P. (2018). Investigating the impacts of increased rural land tenure security: A systematic review of the evidence. *Journal of Rural Studies, 61*, 34–62. https://doi.org/10.1016/j.jrurstud.2018.05.001

Holland, M. B., de Koning, F., Morales, M., Naughton-Treves, L., Robinson, B. E., & Suárez, L. (2014). Complex tenure and deforestation: Implications for conservation incentives in the Ecuadorian Amazon. *World Development, 55*, 21–36. https://doi.org/10.1016/j.worlddev.2013.01.012

Holland, M. B., Jones, K. W., Naughton-Treves, L., Freire, J. L., Morales, M., & Suárez, L. (2017). Titling land to conserve forests: The case of Cuyabeno Reserve in Ecuador. *Global Environmental Change., 44*, 27–38. https://doi.org/10.1016/j.gloenvcha.2017.02.004

Intergovernmental Panel on Climate Change (IPCC). (2020). *Climate Change and Land—Summary for policy makers. Climate Change and Land: an IPCC special report on climate change, desertification, land degradation, sustainable land management, food security, and greenhouse gas fluxes in terrestrial ecosystems.*

Lawry, S., Samii, C., Hall, R., Leopold, A., Hornby, D., & Mtero, F. (2017). The impact of land property rights interventions on investment and agricultural productivity in developing countries: A systematic review. *Journal of Development Effectiveness, 9*(1), 61–81. https://doi.org/10.1080/19439342.2016.1160947

Masuda, Y. J., Kelly, A. C., Robinson, B. E., Holland, M. B., Bedford, C., Childress, M., et al. (2020). How do practitioners characterize land tenure security? *Conservation Science and Practice, 2*(5), e186. https://doi.org/10.1111/csp2.186

McSweeney, K., & Coomes, O. T. (2020). Who owns the Earth? A challenge for the land system science community. *Journal of Land Use Science, 15*, 482–488. https://doi.org/10.1080/1747423X.2020.1765428

Meinzen-Dick, R., Quisumbing, A., Doss, C., & Theis, S. (2019). Women's land rights as a pathway to poverty reduction: Framework and review of available evidence. *Agricultural Systems, 172*, 72–82. https://doi.org/10.1016/j.agsy.2017.10.009

Natcher, D., Hickey, C., Nelson, M., & Davis, S. (2009). Implications of tenure insecurity for aboriginal land use in Canada. *Human Organization, 68*, 245–257. https://doi.org/10.17730/humo.68.3.60pp7583m183t1t1

Naughton-Trevesa, L., Alix-Garcia, J., & Chapman, C. A. (2011). Lessons about parks and poverty from a decade of forest loss and economic growth around Kibale National Park, Uganda. *Proceedings of the National Academy of Sciences of the United States of America, 108*, 13919–13924. https://doi.org/10.1073/pnas.1013332108

Notess, L., Veit, P., Monterroso, I., Andiko, S., Larson, A. M., et al. (2021). Community land formalization and company land acquisition procedures: A review of 33 procedures in 15 countries. *Land Use Policy*, 110, 104461. https://doi.org/10.1016/j.landusepol.2020.104461

O'Sullivan, M. (2017). *Gender and property rights in Sub-Saharan Africa: A review of constraints and effective interventions.* (No. 8250). Policy Research Working Paper. http://documents.worldbank.org/curated/en/326061510086379739/pdf/WPS8239.pdf

Orellano, V., Azevedo, P. F., Saes, M. S., & Nascimento, V. E. (2015). Land invasions, insecure property rights and production decisions. *Journal of Agricultural Economics, 66*, 660–671. https://doi.org/10.1111/1477-9552.12103

Robinson, B. E., Holland, M. B., & Naughton-Treves, L. (2014). Does secure land tenure save forests? A meta-analysis of the relationship between land tenure and tropical deforestation. *Global Environmental Change, 29*, 281–293. https://doi.org/10.1016/j.gloenvcha.2013.05.012

Robinson, B. E., Masuda, Y. J., Kelly, A., Holland, M. B., Bedford, C., Childress, M., et al. (2018). Incorporating land tenure security into conservation. *Conservation Letters, 11*(2), e12383. https://doi.org/10.1111/conl.12383

Shang, Q., & Price, A. (2018). A blockchain-based land titling project for the Republic of Georgia. *Innovations, 12*(3/4), 72–78.

Simbizi, M. C. D., Bennett, R. M., & Zevenbergen, J. (2014). Land tenure security: Revisiting and refining the concept for Sub-Saharan Africa's rural poor. *Land Use Policy, 36*, 231–238. https://doi.org/10.1016/j.landusepol.2013.08.006

Sjaastad, E., & Cousins, B. (2009). Formalisation of land rights in the South: An overview. *Land Use Policy, 26*(1), 1–9. https://doi.org/10.1016/j.landusepol.2008.05.004

Tseng, T.-W. J., Robinson, B. E., Bellemare, M. F., BenYishay, A., Blackman, A., Boucher, T., et al. (2021). Influence of land tenure interventions on human well-being and environmental outcomes. *Nature Sustainability, 4*(3), 242–251. https://doi.org/10.1038/s41893-020-00648-5

United Nations Department of Economic and Social Affairs, P. D. (2019). *World urbanization prospects: The 2018 revision. world urbanization prospects: The 2018 revision*. UN. https://doi.org/10.18356/b9e995fe-en

van Gelder, J. L. (2010). What tenure security? The case for a tripartite view. *Land Use Policy, 27*(2), 449–456. https://doi.org/10.1016/j.landusepol.2009.06.008

Index[1]

[1] Note: Page numbers followed by 'n' refer to notes.

M. B. Holland et al. (eds.), *Land Tenure Security and Sustainable Development*, https://doi.org/10.1007/978-3-030-81881-4

Kolibri 360 GmbH
Ferdinand-Jühlke-Straße 7
99095 Erfurt, Deutschland
produktsicherheit@kolibri360.de